Praise for *What's Math Got to Do with It?*

"Without doubt, this is the most important book parents should read before choosing a school for their child, and teachers should read when considering ways to improve their math teaching. A compelling, readable account of years of research into what works, and what doesn't, in mathematics education."

> —Keith Devlin, PhD, Stanford University mathematician, award-winning author of *The Math Gene* and thirty-one other books, and the Math Guy on NPR

"Jo Boaler vividly shows us—rather than just telling us—what terrific math instruction looks like and, equally important, how sharply it differs from how the subject is usually taught. *What's Math Got to Do with It?* is the first book I recommend to teachers and parents who want to understand the harms of conventional ways of teaching math as well as the benefits of realistic alternatives."

> —Alfie Kohn, author of *The Schools Our Children Deserve* and *Feel-Bad Education*

"There is so much wisdom packed into this engaging little book. Boaler sensibly addresses current hot topics—the Common Core, mind-set, ability grouping, gender differences—but goes way behind them to consider the nature of mathematics itself and offers a wealth of practical advice to parents, teachers, and policy makers. More than ever, we need books on education like his one."

> —Mike Rose, author of *Possible Lives: The Promise of Education in America*

"*What's Math Got to Do with It?* comes the closest of anything that I have read to a manifesto that I would provide to parents to help them better understand the importance of good teaching of interesting and complex mathematics."

> —*Journal for Research in Mathematics Education*

"Jo Boaler shows that math is understandable, and that it can be fun to get your head around it—but that it's often taught in ways that make it dry and deadly. She points to the beauty and joy of mathematics, and ways that math classrooms can become centers of lively mathematical thinking. American children deserve a richer mathematical diet than we've given them, and Boaler shows how and why."

—Alan H. Schoenfeld, Elizabeth and Edward Conner Professor of Education, University of California, Berkeley

"This extraordinary book shows teachers and parents the path to teaching children to enjoy math while they develop deep and flexible understanding. The author practices what she preaches; using systematic research she and others have conducted on two continents, she makes learning about math teaching accessible and fun."

—Deborah Stipek, I. James Quillen Dean and Professor of Education, Stanford University

"Jo Boaler makes a powerful case for a problem-solving approach to teaching mathematics, and she presents the research to back it up. This book should be read by anyone concerned about the education of our children."

—Deborah Schifter, principal research scientist, Education Development Center, Inc.

"Jo Boaler explains with insight and clarity why so many students dislike mathematics and what the rest of us can do about it. Her solutions are comprehensive, grounded in research, and powerfully applied by parents, teachers, and anyone else with an interest in mathematics."

—Dan Meyer, Apple Distinguished Educator and one of *Tech & Learning*'s 30 Leaders of the Future

PENGUIN BOOKS

WHAT'S MATH GOT TO DO WITH IT?

Jo Boaler is a professor of mathematics education at Stanford University and the cofounder of YouCubed. She is also the editor of the Research Commentary section of *The Journal for Research in Mathematics Education* (JRME) and author of the first Massive Open Online Course (MOOC) on mathematics teaching and learning. Former roles have included being the Marie Curie Professor of Mathematics Education in England. She is the recipient of several awards, an adviser to Silicon Valley companies, and a White House presenter on girls and STEM (science, technology, engineering, and mathematics) education. She is a regular contributor to national television and radio in the United States and the UK. Her research has appeared in newspapers around the world including the *Wall Street Journal*, the *Times* (London), and the *Telegraph* (UK).

Monkey Business Images

What's Math Got to Do with It?

How Teachers and Parents Can Transform Mathematics Learning and Inspire Success

Jo Boaler

PENGUIN BOOKS

PENGUIN BOOKS

Published by the Penguin Group
Penguin Group (USA) LLC
375 Hudson Street
New York, New York 10014

USA | Canada | UK | Ireland | Australia | New Zealand | India | South Africa | China
penguin.com
A Penguin Random House Company

First published in the United States of America by Viking Penguin,
a member of Penguin Group (USA) Inc., 2008
Published in Penguin Books 2009
This updated and revised edition published 2015

Photographs by the author unless otherwise indicated.

ISBN 978-0-14-312829-8

Printed in the United States of America
1 0

Set in ITC Garamond Std Light

For all the teachers who have inspired my work

Acknowledgments

This book has been a journey of opportunities. Over recent years I have been able to learn from some of America's most inspirational teachers and their students, and to work alongside visionary friends and colleagues who have broadened and enriched my thinking. I am deeply grateful to many people in California, particularly at Stanford University and in Bay Area schools, who made this book possible.

This book was conceived at a very special place: the Center for Advanced Study in the Behavioral Sciences in California, a place devoted to the generation of ideas. I had given a presentation to the other fellows at the center, a group of scholars who worked in different areas of social science research, on the results of my studies of mathematics learning. The group responded strongly, with expressions of shock and dismay, and they urged me to get my results out to the general public. They convinced me to write a book proposal for a broader audience and many people—in particular Susan Shirk, Sam Popkin, and David Clark—supported me along the way.

From that point I was greatly encouraged by my agent Jill Marsal and Kathryn Court, of Penguin Books, both of whom believed in the book, which meant a lot to me. I wrote the book in the stimulating environment of Stanford's Graduate School of Education, surrounded by a group of graduate students who served as critics and supporters. I would personally like to thank all of my doctoral students, past and present, who contributed to the mathematics education group at Stanford.

I have learned a great deal from some truly inspirational teachers in recent years—among them Cathy Humphreys, Carlos Cabana, Sandie Gilliam, Estelle Woodbury, and Ruth Parker. They change students' lives on a daily basis and I am privileged to have been able to work with them and learn from them. Cathy is a good friend who has helped me in many ways. I am also deeply grateful to the students of Railside, Greendale, Hilltop, Amber Hill, and Phoenix Park schools; they all gave me their honest and insightful feedback on their mathematics learning experiences and they are the reason that I wrote this book.

I am fortunate to have had some great teachers of my own in my life—including Professors Paul Black and Dylan Wiliam, both of whom encouraged me in important ways at an early point in my academic career and kindly read chapters of this book for me. Professor Leone Burton, one of my strongest supporters, will be greatly missed by many people. Most of all I would like to thank my two daughters, Jaime and Ariane, for putting up with me when I lock myself away to write!

Contents

Preface to the New Edition

Introduction

In 2008, when *What's Math Got to Do with It?* was first published, the United States was in the grip of endless multiple-choice testing and widespread math failure, brain research was in its infancy, and a group of traditional mathematicians was working tirelessly to stop school reforms. Fast-forward to 2015 and the landscape has changed with the emergence of incredible new research on the brain and learning, which is being acknowledged and acted upon. The White House has convened numerous meetings over the past few years where researchers, myself included, have talked about mathematics, mind-set, and equity. The traditional mathematicians have lost their voice, and many more people are receptive to the idea of a future in which all children can learn mathematics to high levels. These changes pave the way for the creation of mathematics classrooms in which students are excited to learn and teachers are armed with the most important knowledge that inspires students to achieve excellence in math.

I love books. I enjoy reading and writing them, but the Internet has helped me to achieve something that is very important to my long-term professional mission. In the summer of 2013, I tried an experiment. A few months earlier I had been introduced to Sebastian Thrun, an amazing man who invented self-driving cars, led Google teams developing Google Maps and Google Glass, and is the CEO of Udacity. The world of free online courses, or MOOCs (massive open online courses), is generally attributed to Sebastian and his colleague Peter Norvig, who is the director of research at Google. Sebastian and Peter decided to put one of Sebastian's Stanford computer science courses online. One hundred sixty thousand people registered for the course, and the MOOC world began. Sebastian went on to create Udacity, an online-course provider, and some months later asked me to help with course design. The time I spent at Udacity was enough to give me the knowledge I needed to design my own course. Putting knowledge about the brain and math learning into the hands of our nation's teachers and parents would change everything and lent itself perfectly to online courses that could share information widely. I designed my courses to make them engaging and interactive, but the levels of interest in the classes surpassed my greatest hopes. To date more than 130,000 people—teachers, parents, and students—have taken my online courses entitled "How to Learn Math," and they are now armed with the critical information that I will share in this book.

I took Udacity's idea that online courses should not be talking heads and adopted their principle that professors should not talk for more than two minutes before engaging the learners in a task. I didn't have much time, with a full-time job teaching at Stanford, researching, and looking after a large team of doctoral students, but I spent every spare moment on weekends and in the evenings creating my experimental course. The course appeared on Stanford's online platform in the summer of 2013 but was not advertised in any way. The course opened with approximately five thousand registrants, but word of mouth spread quickly, and by the time the course finished at the end of the summer more than forty thousand teachers and parents had enrolled. Many MOOCs have high enrollment, but generally only a small proportion of the people who register end up taking the courses. This was not the case with my course, and an impressive 63 percent of people completed most of the course. Even more rewardingly, at the end of the course 95 percent of teachers and parents said that they would change their teaching/ parenting as a result of the ideas they had learned in the class. In the months after my class, news of the ideas spread, hundreds of videos were posted to YouTube, my in-box was flooded, and requests for speaking engagements increased. So many teachers and parents asked for continued access to the new ideas that Cathy Williams and I launched YouCubed. YouCubed initially was a nonprofit company but now is a center at Stanford. In the summer of 2014, I published a new online course getting the same powerful ideas straight to students, and within a few months of the course opening eighty-five thousand students had taken the class or been shown the videos by their teachers.

I do not think that online courses are the most powerful medium for learning—I would always rather interact with groups of learners face-to-face and have them discuss ideas with each other—but online courses allow wide-scale access to important knowledge that urgently needs to be shared and that cannot depend on parents finding their way to high-quality teaching in their local area. I have been frustrated over the years by universities encouraging professors to publish research on learning only in academic journals, which are read by other academics and do not get to the people who need them—teachers and parents. Important knowledge on ways to learn effectively is usually locked away in journals and libraries unless researchers choose to publish their findings in different, more accessible ways and universities do not penalize them for doing so. This is what motivated me to publish the first edition of *What's Math Got to Do with It?* and to update this edition with the latest research and practical information on ways to help all learners of mathematics.

When I wrote the original edition of *What's Math Got to Do with It?* the field of mathematics education had a large body of research on ways to teach and learn mathematics well. We knew how to empower math learners, but the research was not getting to teachers or being used in classrooms. If you walk into most math classrooms in the United States, particularly at the high school level, you would think you had been transported into the Victorian age. For the most part teachers are still at the front of the room lecturing on methods, students are still at desks learning to calculate by hand, and the mathematics being taught is three-hundred-year-old mathematics that is not needed in the modern world.[1] In elementary classrooms across the United States, students are turned away from mathematics on a daily basis by timed tests and speed competitions, which we know can cause the early onset of math anxiety for many students.[2, 3, 4] We have the research knowledge to change this and for classrooms to become places where all students are inspired by mathematics. In the last decade, important new research on the brain and learning has emerged, which is critical for math teachers, math learners, and parents everywhere.

Mind-set and Mathematics

In 2006, a trade book appeared on bookshelves that ultimately would have one of the biggest impacts of any research volume ever published in education. In *Mindset: The New Psychology of Success*, Stanford Professor Carol Dweck summarized key findings from her research on the nature and impact of mind-sets. The book quickly became a *New York Times* bestseller and was translated into more than twenty languages. Dweck's decades of research with people of various ages showed that students with a "growth mind-set"—who believe that intelligence and "smartness" can be learned—go on to higher levels of achievement, engagement, and persistence. The implications of this mind-set are profound, especially for students of mathematics.

When I returned to Stanford in 2010, one of the first things I did was arrange to meet Carol Dweck. She agreed with me that mathematics is the subject most in need of a mind-set makeover and that mathematics teachers are the group who could benefit the most with knowledge of mind-set. Since that meeting we have been working together, writing, researching, and collaborating with teachers.

Mathematics, more than any other subject, has the power to crush students' confidence. The reasons are related both to the teaching methods that prevail in US math classrooms and the fixed ideas about mathematics held by the majority of the US population and passed on to our children from birth. One of the most damaging mathematics myths propagated in classrooms and homes is that math is a gift—that some people are naturally good at math and some are not.[5, 6] This idea is strangely cherished in the Western world but virtually absent in Eastern countries such as China and Japan that top the world in mathematics achievement.[7]

New scientific evidence showing the incredible capacity of the brain to change, rewire, and grow in a really short time[8] tells us that all students can learn mathematics to high levels with good teaching experiences. Traditional educators believe that some students do not have the brains to be able to work on complex mathematics, but it is working on complex mathematics that enables brain connections to develop. Students can grasp high-level ideas, but they will not develop the brain connections that allow them to do so if they are given low-level work and negative messages about their own potential.[9]

As I work with schools and districts encouraging mathematics

mathe learning
≠ performance (testing)

teaching that promotes growth rather than fixed mind-sets (see www.youcubed.org), a critical requirement is that teachers offer mathematics as a learning subject, not a performance subject. Most students when asked what they think their role is in math classrooms say it is to answer questions correctly. They don't think they are in math classrooms to appreciate the beauty of mathematics, to explore the rich set of connections that make up the subject, or even to learn about the applicability of the subject. They think they are in math classrooms to perform. This was brought home to me recently when a colleague, Rachel Lambert, told me her six-year-old son had come home saying he didn't like math. When she asked him why, he said, "Math is too much answer time and not enough learning time." Students from kindergarten upward realize that math is different from other subjects: learning gives way to answering questions and taking tests—performing.

For students to see mathematics as a subject of learning, they need tasks and questions in math class that allow for learning. Chapter 3 will show some of these tasks and how they are used in classrooms. Teachers need to stop giving the wrong messages to students, whether through grouping, grading, or short, narrow math problems, which themselves imply that math is a question of the right or wrong answer rather than a learning subject. In chapters 4 and 5, I will describe the highly productive ways that teachers can grade, assess, and group students. The students who suffer most from fixed-mind-set thinking are high-performing girls, and chapter 6 will explain more about this phenomenon and ways we can give girls a positive and equitable future. Chapter 7 is focused on the important role parents can play and provides suggestions for activities and advice that can be used in the home.

One of the most interesting findings from research on the brain to emerge over recent years is something that I try to communicate as widely as I can. We now know that when students

[handwritten: mistakes = learning + growth]
[handwritten: → growth: challenging + enaging problems]

make a mistake in math, their brain grows, synapses fire, and connections are made.[10] This finding tells us that we want students to make mistakes in math class and that students should not view mistakes as learning failures but as learning achievements.[11] But students everywhere feel terrible when they make a mistake. They think it means they are not a "math person." We need to change this thinking by telling students that mistakes are productive. When I talk to teachers about this research they often say, "But surely students have to work through their error and see why it is a mistake for brains to grow." This is a reasonable assumption, but students do not even need to know they have made a mistake for brains to grow. What research tells us is that when a mistake is made there are two potential brain sparks: the first one comes when we make a mistake but are not aware of the mistake; the second comes when we realize we have made a mistake. How can this be? How can our brains grow when we do not even know we have made a mistake? The best knowledge we have on this question tells us that our brains grow when we make mistakes because those are times of struggle, and our brains grow the most when we are challenged and engaging with difficult, conceptual questions.

Carol Dweck and I sometimes present together in workshops for teachers and parents. One of the pieces of advice Carol gives to parents in our workshops is that when children come home from school and say that they got all of their work correct that day, parents should say, "Oh, I am sorry, then you didn't get the opportunity to learn today." She is making a good point, and we need to shift teachers' and students' thinking about what they should aim for in mathematics lessons. Teachers care deeply about their students, and it is typical to arrange math lessons so that students are getting most of their work correct. This makes students feel good, but it is not the most productive learning environment for them. We need to change math lessons to make them challenging for all students, and we *comfortable making mistakes*

need to change students' mind-sets so that they know <u>it is pro-ductive to struggle and make mistakes and that they should feel comfortable doing so.</u>

While I was sitting in an elementary classroom in Shanghai recently, the principal leaned over to tell me that the teacher was calling on students who had made mistakes to share with the whole class so that they could all learn. The students seemed pleased to be given the opportunity to share their incorrect thinking. Instead of mathematics classroom lessons filled with short questions that students are intended to get right or wrong, they need to be filled with open-ended tasks that include space for learning as well as space for struggle and growth. YouCubed provides examples of the most productive tasks students should work on at home and in school.

Mathematics and the Common Core

The new Common Core mathematics standards (www.core standards.org) have prompted considerable controversy across the United States. Most of the opponents of the Common Core are politically motivated. Some oppose the Common Core because of the tests that are being written to assess the new standards, but opposing the Common Core because of the standardized tests is somewhat misguided since the tests are not part of the Common Core and really are a separate policy decision, which should be considered outside of the curriculum. Others oppose the Common Core curriculum because their children are successful in the traditional model of math teaching, and they want to keep that advantage. The most curious opposition comes from parents who say that the math in the Common Core is too hard for their children. So what do we know about the Common Core, from research? And is its introduction in the United States a good or a bad thing?

The Common Core math curriculum is not the curriculum I

would have designed if I had had the chance. It still has far too much content that is not relevant for the modern world and that turns students off mathematics, particularly in the high school years, but it is a step in the right direction—for a number of reasons. The most important improvement is the inclusion of a set of standards called the "mathematical practices." The practice standards do not set out knowledge to be learned, as the other standards do, but ways of *being mathematical*. They describe aspects of mathematics such as problem solving, sense making, persevering, and reasoning. It is critically important that students work in these ways; these actions have been part of curricula in other countries for many years. Now that these methods of thinking are part of the US curriculum, students should be spending time in classrooms using mathematics in these ways.

The inclusion of these practices means that the tasks students work on will change. Students will be given more challenging tasks and spoon-fed less. Instead of being told a method and then practicing it, they will need to learn to choose, adapt, and use methods. Students need to learn to problem solve and persist when tasks are longer or more challenging. This is really important work for teachers and students. My main problem with the Common Core standards is that they require teachers to engage students in thinking about what makes sense and problem solving; but such activities, which involve going deeper into the mathematics, take more time. In the elementary and middle grades, the content has been reduced so that teachers and students can take the time to work in these productive ways, but the high school standards are as packed as ever with obsolete content that works against teachers being able to go into depth and give students the experiences they need. For more information on the potential impact of the Common Core go to http://youcubed.org/parents/2014/why-we-need-common-core-math.

Many people think that countries such as China, who top the world in mathematics achievement, do so by drilling students in content, but this is far from the truth. In Shanghai, the highest scoring region of China, I watched numerous high school lessons, and in no lesson did the teachers work on more than three questions in an hour. What was staggering was the depth to which teachers and students delved into each question, exploring every aspect of the mathematics. In all of the lessons, the students talked more than the teacher as they discussed what they were learning. We recorded one of the lessons, and it can be seen on YouCubed. This is a model of mathematics teaching that we need in the United States.

Currently, more than half of all US students fail mathematics, and mathematics is a harshly inequitable subject.[12, 13] When our classrooms change—when students are encouraged to believe that they can be successful in mathematics and are taught using the high-quality teaching methods we know work—the landscape of mathematics teaching and learning in the United States will change forever. This book gives readers—whether teachers, administrators, or parents—the knowledge needed to make the important changes required for our students in the United States and for the future of our society. I hope you enjoy it, whether you are a new reader or one of the many who read the original *What's Math Got to Do with It?*, and are looking to be reenergized by the new ideas shared within this updated edition.

What's Math
Got to Do with It?

Zurijeta

Introduction

Understanding the Urgency

A number of years ago in California, I attended a math lesson that I will never forget. Several people had recommended that I visit the class to see an amazing and unusual teacher, and as I climbed the steps of the classroom I was excited. I knocked on the door. Nobody answered, so I pushed open the door and stepped inside. Emily Moskam's class was not as quiet as most math classes I have visited. A group of tall adolescent boys stood at the front, smiling and laughing, as they worked on a math problem. One of the boys spoke excitedly as he jumped around explaining his ideas. Sunlight streamed through the windows, giving the front of the room a stagelike quality. I moved quietly among the rows of students to take a seat at the side of the room.

In recognition of my arrival, Emily nodded briefly in my direction. All eyes were on the front, and I realized that the students had not heard the door because they were deeply engaged in a problem Emily had sketched on the board. They were working out the time it would take a skateboarder to crash into a padded wall after holding on to a spinning merry-go-round and letting go at a certain point. The problem was complicated, involving high-level mathematics. Nobody had the solution, but various students were offering ideas. After the boys sat down, three girls went to the board and added to the boys' work, taking their ideas further. Ryan, a tall boy sporting a large football championship ring, was sitting at the back and he asked them, "What was your goal for the end product?" The three girls explained that they were first finding the rate that the skateboarder was traveling. After that, they would try to find the distance from the merry-go-round to the wall. From there things moved quickly and animatedly in the class. Different students went to the board, sometimes in pairs or groups, sometimes alone, to share their ideas. Within ten minutes, the class had solved the problem by drawing from trigonometry and geometry, using similar triangles and tangent lines. The students had worked together like a well-oiled machine, connecting different mathematical ideas as they worked toward a solution. The math was hard and I was impressed. (The full question and the solution for this math problem, as well as the other puzzles in this book, are in appendix A.)

Unusually for a math class it was the students, not the teacher, who had solved the problem. Most students in the class had contributed something, and they had been excited about their work. As the students shared ideas, others listened carefully and built upon them. Heated debates have raged between those who believe that mathematics should be taught traditionally—with the teacher explaining methods and the students watching and then practicing them, in silence—and those who believe that students should be more involved—discussing ideas and solving applied problems. Those in the "traditional" camp have worried that student-centered teaching approaches sacrifice standard methods, mathematical correctness, or high-level work. But this class was a perfect example of one that would please people on both sides of the debate, as the students fluently made use of high-level mathematics, which they applied with precision. At the same time, the students were actively involved in their learning and were able to offer their own thoughts in solving problems. This class worked so well because students were given problems that interested and challenged them and, also, they were allowed to spend part of each lesson working alone and part of each lesson talking with one another and sharing ideas about math. As the students filed out of the room at the end of class, one of the boys sighed, "I love this class." His friend agreed.

Unfortunately, very few math classes are like Emily Moskam's, and their scarcity is part of the problem with American math education. Instead of actively engaging students in mathematical problem solving, most American math classes have them sitting in rows and listening to a teacher demonstrate methods that students neither understand nor care about. Far too many students in America *hate* math and for many it is a source of anxiety and fear. American students do not achieve well and they do not choose to study mathematics beyond basic courses, a situation that presents serious risks to the future medical, scientific, and technological advancement of society. Consider, for example, these chilling facts:

- In a recent international assessment of mathematics performance conducted in sixty-four countries across the world, the United States ranked a lowly thirty-six.[1] When levels of spending in education were taken into account, the United States dropped to the bottom of all the tested countries.
- Currently, only 1 percent of undergraduates major in math,[2] and the last ten years have shown a 5 percent decline in women math majors.[3]

- In a survey of middle school students, more than half said they would rather eat broccoli than do math, and 44 percent would rather take out the trash.[4]

- Approximately 50 percent of students in the United States attend two-year colleges. About 70 percent of those students are placed into remedial math courses repeating the math they took in high school. Only one in ten of the students passes the course. The rest leave or fail.[5] For about fifteen million students in the United States, math ends their college career.

In September 1989, the nation's governors gathered in Charlottesville, Virginia, and set a challenge for the new millennium: American children should top the world in mathematics and science by the year 2000. Now, fifteen years after that time goal, the United States sits near the bottom of international rankings of mathematics achievement.

Achievement and interest among children is low, but the problem does not stop there. Mathematics is widely hated among adults because of their school experiences, and most adults avoid mathematics at all costs. Yet the advent of new technologies means that all adults now need to be able to reason mathematically in order to work and live in today's society. What's more, mathematics could be a source of great interest and pleasure for Americans, if only they could forget their past experiences and see mathematics for what it is rather than the distorted image that was presented to them in school. When I tell people that I am a professor of mathematics education, they often shriek in horror, saying that they cannot do math to save their lives. This always makes me sad because I know that they must have experienced bad math teaching. I recently interviewed a group of young adults who had hated math in school and were surprised at how interesting math was in their work;

many of them had even started working on math puzzles in their spare time. They could not understand why school had misrepresented the subject so badly.

This aversion to math is reflected in our popular culture as well: in an episode of *The Simpsons,* Bart Simpson returns school textbooks to his teacher at the end of the year, noting that they are all in perfect shape—and in the case of his math book, "still in their original wrapping!" Bart's disinclination to open his math book probably resonated with many of his school-age viewers. Barbie's first words may have done the same with her young owners. When she finally started speaking, her first words were "Math class is tough"—a feature that the manufacturers quickly and correctly withdrew. But Barbie and Bart are not alone—an Associated Press–America Online (AOL) news poll showed that a staggering four out of ten adults said that they *hated* math in school. Twice as many people hated math as any other subject.

[handwritten margin note: 40 % adults hate math]

Then again, amid this picture of widespread disdain there is evidence that mathematics may have the potential to be quite appealing. The mathematical TV show *NUMB3RS* developed a cult following during its first season. Sudoku, the ancient Japanese number puzzle, has gripped America. Sudoku involves filling nine 3 × 3 squares so that the numbers 1 to 9 appear only once in each row or column. Americans everywhere can be seen hunched over their number grids before, during, and after work, engaging in the most mathematical of acts—logical thinking. These trends suggest something interesting: school math is widely hated, but the mathematics of life, work, and leisure is intriguing and much more enjoyable. There are two versions of math in the lives of many Americans: the strange and boring subject that they encountered in classrooms, and an interesting set of ideas that is the math of the world and is curiously different and surprisingly engaging. Our task is to introduce this second version to today's students, get them excited about math, and prepare them for the future.

The Mathematics of Work and Life

What sort of mathematics will young people need in the future? Ray Peacock, a respected employer who was the research director of Phillips Laboratories in the UK, reflected upon the qualities needed in the high-tech workplace:

> Lots of people think knowledge is what we want, and I don't believe that, because knowledge is astonishingly transitory. We don't employ people as knowledge bases, we employ people to actually do things or solve things. . . . Knowledge bases come out of books. So I want flexibility and continuous learning. . . . [A]nd I need teamworking. And part of teamworking is communications. . . . When you are out doing any job, in any business . . . the tasks are not 45 minutes max, they're usually 3-week dollops or one-day dollops, or something, and the guy who gives up, oh sod it, you don't want him. So the things therefore are the flexibility, the teamworking, communications, and the sheer persistence. *Future job characteristics*

Dr. Peacock is not alone in valuing problem solving and flexibility. Surveys of American employers from manufacturing, information technology, and the skilled trades tell us that employers want young people who can use "statistics and three-dimensional geometry, systems thinking, and estimation skills. Even more important, they need the disposition to think through problems that blend quantitative work with verbal, visual, and mechanical information; the capacity to interpret and present technical information; and the ability to deal with situations when something goes wrong."[6]

The mathematics that people need is not the sort of math learned in most classrooms. People do not need to regurgitate hundreds of standard methods. They need to reason and problem solve, flexibly applying methods in new situations. Mathematics is now so critical to American citizens that some have labeled it the "new civil right."[7] If young people are to become powerful citizens with full control over their lives, then they need to be able to reason mathematically—to think logically, compare numbers, analyze evidence, and reason with numbers.[8] *Businessweek* declared that "the world is moving into a new age of numbers." Mathematics classrooms need to catch up—not only to help future employers and employees, or even to give students a taste of authentic mathematics, but to prepare young people for their lives.

Engineering is one of the most mathematical of jobs, and entrants to the profession need to be proficient in high levels of math in order to be considered. Julie Gainsburg studied structural engineers at work for more than seventy hours and found that although they used mathematics extensively in their work, they rarely used standard methods and procedures. Typically the engineers needed to interpret the problems they were asked to solve (such as the design of a parking lot or the support of a wall) and form a simplified model to which they could apply mathematical methods. They would then select and adapt methods that could be applied to their models, run calculations (using various representations—graphs, words, equations, pictures, and tables—as they worked), and justify and communicate their methods and results. The engineers engaged in flexible problem solving, adapting and using mathematics. Although they occasionally faced situations when they could simply use standard mathematical formulas, this was rare and the problems they worked on were "usually ill-structured and open-ended." As Gainsburg writes, "Recognizing and defining the problem and wrangling it into a solvable shape are often part of the work;

[handwritten margin note: identifying a problem and how to solve it is a part of the effort — which method]

methods for solving have to be chosen, or adapted from multiple possibilities, or even invented; multiple solutions are usually possible; and identifying the 'best' route is rarely a clear-cut determination."[9] Gainsburg's findings echo those of other studies of high-level mathematics in use in such areas as design, technology, and medicine.[10, 11] Gainsburg's conclusion is definitive and damning: "The traditional K–12 mathematics curriculum, with its focus on performing computational manipulations, is unlikely to prepare students for the problem-solving demands of the high-tech workplace."[12]

Conrad Wolfram agrees. He is one of the directors of Wolfram/ Alpha, one of the biggest mathematical companies in the world, which also owns Mathematica. Conrad is heading a movement to stop mathematics from being reduced to calculating. In his TED talk watched by more than one million people, he describes mathematical work as a four-step process: first posing a question, then constructing a model to help answer the question, next performing a calculation, and finally converting the model back to the real-world situation by thinking about whether it answered the question (see http://www.ted.com /talks/conrad_wolfram_teaching_kids_real_math_with _computers).

[handwritten margin note: Less time teaching calculations / more asking questions + set up models]

He then makes an important point. Teachers spend 80 percent of the time teaching students the calculation step when they should be using computers for that part and spending more time helping students to ask questions and to set up and interpret models. Sebastian Thrun, the CEO of Udacity, appears in my online course making an important point for teachers and parents: we do not and cannot know what mathematics students will need in the future, but the best preparation we can give them is to teach them to be quantitatively literate, think flexibly and creatively, problem solve, and use intuition as they develop mathematical ideas.[13] The chapters that follow will explain and illustrate how this happens in classrooms.

Studies of people using mathematics in their lives—when shopping and performing other routine tasks—lead to similar recommendations. Researchers have found that adults cope well with mathematical demands, but they draw from school knowledge infrequently. In real-world situations, such as in street markets and shops, individuals have rarely made use of any school-learned mathematical methods or procedures. Instead, they have created methods that work given the constraints of the situations they faced.[14, 15, 16, 17] Jean Lave, a professor at the University of California, Berkeley, found that shoppers used their own methods to work out which were better deals in shops, without using any formal methods learned in school, and that dieters used informal methods that they created when needing to work out measures of servings. For example, a dieter who was told he could eat 3/4 of a 2/3 cup of cottage cheese did not perform the standard algorithm for multiplying the fractions. Instead, he emptied 2/3 of a cup onto a measuring board, patted it into a circle, marked a cross on it, and took one quadrant away, leaving 3/4 of it.[18] Lave gives many examples from her various studies of people using informal methods such as these, without any recourse to school-learned methods.

example of math by dieter

The ways in which people use mathematics in the world will probably sound familiar to most readers, as they are the ways many of us use mathematics. Adults rarely stop to remember formal algorithms; instead, successful math users size up situations and adapt and apply mathematical methods, using them flexibly.

The math of the world is so different from the math taught in most classrooms that young people often leave school ill-prepared for the demands of their work and lives. Children learn, even when they are still in school, about the irrelevance of the work they are given, an issue that becomes increasingly important to children as they move through adolescence. As part of a research study in England,[19] I interviewed students who had learned traditionally, as well as those who had learned through a problem-solving approach, about their use of math in their part-time after-school jobs. The students from the traditional approach all said that they used and needed math out of school but that they would never, ever make use of the mathematics they were learning in their school classrooms. The students regarded the school math classroom as a separate world with clear boundaries that separated it from their lives. The students who had learned through a problem-solving approach did not regard the mathematics of school and the world as different and talked with ease about their use of the school-learned math in their jobs and lives.

When students from sixty-four countries were given tests of mathematical problem solving in 2012, the United States came in thirty-sixth place. This is despite the fact that problem-solving approaches not only teach young people how to solve problems using mathematics but also prepare them for examinations—as well as or better than traditional approaches.[20, 21]

What's math got to do with it? It has a lot to do with children having low self-esteem throughout their lives because they are made to feel bad in math classes; it also has a lot to do with children not enjoying school as they are made to sit through uninspiring lessons, and it has a lot to do with the future of the country, given that we urgently need many more mathematical people to help with jobs in science, medicine, technology, and other fields. Math has many things to answer for and this book is all about giving teachers, parents, and others the knowledge of good ways to work in schools and homes, so that we can start improving our children's and our country's futures.

We need to bring *mathematics* back into math classrooms and children's lives, and we must treat this as a matter of urgency. The classroom characteristics that I am arguing for in this book are not at either of the poles of a "traditional" or "reform" debate, and they could take place in any math classroom or home because they are all about *being mathematical*. Children need to solve complex problems, to ask many forms of questions, and to use, adapt, and apply standard methods, as well as to make connections between methods and to reason mathematically—and they can engage in such methods at home and at school.

But let us return for a moment to Emily Moskam's classroom, the one that I described at the start of this chapter. The class was in a public high school that offered students a choice between a "traditional" and a problem-solving approach. When I took a senior professor from Stanford to visit Emily's class, he simply described it as "magical." Perhaps his enthusiasm was not surprising. Emily Moskam had won awards for her teaching, including a Presidential Award, and her students regularly went on to pursue mathematical careers. What is surprising and *tragic* is that soon after I videotaped this lesson, Emily was told that she could no longer teach in this way. A small group of parents had worked hard to convince others at the school that all students needed to learn mathematics using only traditional methods—that students should sit in rows and not be asked to solve complex problems and only the teacher should talk. From that year on, math classes at Emily's school[22, 23] looked very different. In some ways they were indistinguishable from classrooms in the 1950s. Teachers stood it the front, explaining procedures to students who sat and quietly practiced them individually. The problem-solving approach that Emily had shown to be so successful for her students is no longer an option in her school.

How Can This Book Help?

I conduct longitudinal studies of children's mathematics learning. These studies are very rare. Typically researchers visit classrooms at a particular point in time to observe children learning, but I have followed thousands of American and British students through years of middle and high school math classes to observe how their learning develops over time. I am a professor of mathematics education at Stanford University and the cofounder of YouCubed, a center I formed to get free and accessible research ideas to teachers and parents. I was formerly the Marie Curie Professor for Mathematics Education in Europe and prior to that a middle and high school math teacher.

In my studies I monitor how students are learning, finding out what's helpful to them and what's not. In recent years, I returned to a middle school in the Bay Area to teach school math again, with some of my graduate students. We taught classes of largely disaffected students who hated math and were getting D's and F's in it. They started the class saying they didn't want to be there, but they ended up loving it, telling us that it had transformed their view of math. One boy told us that if math was like that during the school year, he would take it all day and every day. One of the girls told us that math class had always appeared so black and white to her, but in this class it was "all the colors in the rainbow." Our teaching methods were not revolutionary: we talked with the children about math, and we worked on algebra and arithmetic through puzzles and problems such as the chessboard problem:

How many squares are there in a chessboard?
(The answer is not 64! See appendix A for details.)

Successful teachers use teaching methods that more people should know about. Good students also use strategies that make them successful—they are not just people who are born with some sort of math gene, as many people think. High-achieving students are people who learn, whether through great teachers, role models, family, or other sources, to use the particular strategies that I will share in this book.

Based on my studies of thousands of children, *What's Math*

Got to Do with It? will identify the problems that American students encounter, and it will share some solutions. I know that many parents and some teachers are afraid of math and don't think that they know enough to help children or even to talk with them about math, especially when high school courses begin. But parents and teachers, whatever their level of math anxiety, can change everything for students by giving positive messages and sharing helpful math strategies that I provide in this book. I hope that this book will spark an interest in people who have been wounded by math experiences, motivate those who already know and enjoy math, and guide teachers and parents to inspire children through mathematical exploration and connection making.

sl_photo

1 / What Is Math?

And Why Do We All Need It?

In my different research studies I have asked hundreds of children, taught traditionally, to tell me what math is. They will typically say such things as "numbers" or "lots of rules." Ask mathematicians what math is and they will more typically tell you that it is "the study of patterns" or "a set of connected ideas." Students of other subjects, such as English and science, give descriptions of their subjects that are similar to those of professors in the same fields. Why is math so different? And why is it that students of math develop such a distorted view of the subject?

Reuben Hersh, a philosopher and mathematician, has written a book called *What Is Mathematics, Really?* in which he explores the true nature of mathematics and makes an important point: people don't like mathematics because of the way it is *misrepresented* in school. The math that millions of Americans experience in school is an impoverished version of the subject and it bears little resemblance to the mathematics of life or work or even the mathematics in which mathematicians engage.

What Is Mathematics, Really?

Mathematics is a human activity, a social phenomenon, a set of methods used to help illuminate the world, and it is part of our culture. In Dan Brown's bestselling novel *The Da Vinci Code*,[1] the author introduces readers to the "divine proportion," a ratio that is also known as the Greek letter phi. This ratio was first discovered in 1202 when Leonardo Pisano, better known as Fibonacci, asked a question about the mating behavior of rabbits. He posed this problem:

> A certain man put a pair of rabbits in a place surrounded on all sides by a wall. How many pairs of rabbits can be produced from that pair in a year if it is supposed that every month each pair begets a new pair which from the second month on becomes productive?

The resulting sequence of pairs of rabbits, now known as the Fibonacci sequence, is

$$1, 1, 2, 3, 5, 8, 13, \ldots$$

What is fascinating about this number pattern is that after the sequence starts, any number can be found from the sum of the previous two numbers. Even more interesting, as we move along the sequence of numbers, dividing each number by the one before it, produces a ratio that gets closer and closer to 1.618, also known as phi or *the golden ratio*. What is amazing about this ratio is that it exists throughout nature. When flower seeds grow in spirals, they grow in the ratio 1.618:1. The ratio of spirals in seashells, pinecones, and pineapples is exactly the same. For example, if you look very carefully at the photograph of a daisy here, you will see that the seeds in the center of the flower form spirals, some of which curve to the left and some to the right.

If you map the spirals carefully, you will see that close to the center there are twenty-one running counterclockwise. Just a little further out, there are thirty-four spirals running clockwise. These numbers appear next to each other in the Fibonacci sequence.

Daisy showing twenty-one
counterclockwise spirals

Daisy showing thirty-four
clockwise spirals

Remarkably, the measurements of various parts of the human body have the exact same relationship. Examples include a person's height divided by the distance from belly button to the floor; and the distance from shoulders to fingertips, divided by the distance from elbows to fingertips. The ratio turns out to be so pleasing to the eye that it is also ubiquitous in art and architecture, even being featured in the United Nations building, the Parthenon in Athens, and the pyramids of Egypt.

Ask most mathematics students in middle or high school about these relationships, and they will not even know they exist. This is not their fault, of course. They have never been taught about them. Mathematics is all about illuminating relationships such as those found in shapes and in nature. It is also a powerful way of expressing relationships and ideas in numerical, graphical, symbolic, verbal, and pictorial forms. This is the wonder of mathematics that is denied to most children.

Those children who do learn about the true nature of mathematics are very fortunate, and it often shapes their lives. Margaret Wertheim, a science reporter for the *New York Times*, reflects upon an Australian mathematics classroom from her childhood and the way that it changed her view of the world:

> When I was ten years old I had what I can only describe as a mystical experience. It came during a math class. We were learning about circles, and to his eternal credit our teacher, Mr. Marshall, let us discover for ourselves the secret image of this unique shape: the number known as pi. Almost everything you want to say about circles can be said in terms of pi, and it seemed to me in my childhood innocence that a great treasure of the universe had just been revealed. Everywhere I looked I saw circles, and at the heart of every one of them was this mysterious number. It was in the shape of the sun and the moon and the earth; in mushrooms, sunflowers, oranges, and pearls;

in wheels, clock faces, crockery, and telephone dials. All of these things were united by pi, yet it transcended them all. I was enchanted. It was as if someone had lifted a veil and shown me a glimpse of a marvelous realm beyond the one I experienced with my senses. From that day on I knew I wanted to know more about the mathematical secrets hidden in the world around me.[2]

How many students who have sat through American math classes would describe mathematics in this way? Why are they not enchanted, as Wertheim was, by the wonder of mathematics, the insights it provides into the world, the way it elucidates the patterns and relationships all around us? It is because they are misled by the image of math presented in school mathematics classrooms and they are not given an opportunity to experience real mathematics. Ask most school students what math is and they will tell you it is a list of rules and procedures that need to be remembered.[3] Their descriptions are frequently focused on calculations. Yet as Keith Devlin, mathematician and writer of several books about math, points out, mathematicians are often not even very good at calculations as they do not feature centrally in their work. As I had mentioned before, ask mathematicians what math is and they are more likely to describe it as *the study of patterns*.[4, 5]

Early in his book *The Math Gene*, Devlin tells us that he hated math in elementary school. He then recalls his reading of W. W. Sawyer's book *Prelude to Mathematics* during high school, which captivated his thinking and even made him consider becoming a mathematician himself. One of Devlin's passages begins with a quote from Sawyer's book:

"Mathematics is the classification and study of all possible patterns." *Pattern* is here used in a way that everybody may agree with. It is to be understood in a very wide sense, to cover almost *any kind of regularity that can be recognized by the mind* [emphasis added]. Life, and certainly intellectual life, is only possible because there are certain regularities in the world. A bird recognizes the black and yellow bands of a wasp; man recognizes that the growth of a plant follows the sowing of a seed. In each case, a mind is aware of pattern.[6]

Reading Sawyer's book was a fortunate event for Devlin, but insights into the true nature of mathematics should not be gained *in spite* of school experiences, nor should they be left to the few who stumble upon the writings of mathematicians. I will argue, as others have done before me, that school classrooms should give children a sense of the nature of mathematics, and that such an endeavor is critical in halting the low achievement and participation that extends across America. Schoolchildren know what English literature and science are because they engage in authentic versions of the subjects in school. Why should mathematics be so different?[7]

What Do Mathematicians Do, Really?

Fermat's last theorem, as it came to be known, was a theory proposed by the great French mathematician Pierre de Fermat in the 1630s. Proving (or disproving) the theory that Fermat set out became the challenge for centuries of mathematicians and caused the theory to become known as "the world's greatest mathematical problem."[8] Fermat (1601–1665) was famous in his time for posing intriguing puzzles and discovering interesting relationships between numbers. Fermat claimed that the equation $a^n + b^n = c^n$ has no whole number solutions for n when n is greater than 2. So, for example, no numbers could make the statement $a^3 + b^3 = c^3$ true. Fermat developed his theory through consideration of Pythagoras's famous case of $a^2 + b^2 = c^2$. Schoolchildren are typically introduced to the Pythagorean formula when learning about triangles, as any right-angled triangle has the property that the sum of squares built on the two sides $(a^2 + b^2)$ is equal to the square of the hypotenuse, c^2.

So, for example, when the sides of a triangle are 3 and 4, then the hypotenuse must be 5 because $3^2 + 4^2 = 5^2$. Sets of three numbers that satisfy Pythagoras's case are those where two square numbers (e.g., 4, 9, 16, 25) can combine to produce a third.

Fermat was intrigued by the Pythagorean triples and explored the case of cube numbers, reasonably expecting that some pairs of cubed numbers could be combined to produce a third cube. But Fermat found that this was not the case and the resulting cube always has too few or too many blocks. For example,

A Close Call

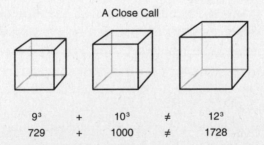

| 9^3 | + | 10^3 | ≠ | 12^3 |
| 729 | + | 1000 | ≠ | 1728 |

The sum of the volumes of cubes of dimensions 9 and 10 almost equals the volume of a cube of dimension 12, but not quite. (It is one short!)

FERMAT'S THEOREM

Indeed, Fermat went on to claim that even if every number in the world was tried, no one would ever find a solution to $a^3 + b^3 = c^3$ nor to $a^4 + b^4 = c^4$, or to any higher power. This was a bold claim involving the universe of numbers. In mathematics it is not enough to make such claims, even if the claims are backed up by hundreds of cases, as mathematics is all about the construction of time-resistant proofs. Mathematical proofs involve making a series of logical statements from which only one conclusion can follow and, once constructed, they are always true. Fermat made an important claim in the 1630s, but he did not provide a proof—and it was the proof of his claim that would elude and frustrate mathematicians for more than 350 years. Not only did Fermat not provide a proof, but he scribbled a note in the margin of his work saying that he had a "marvelous" proof of his claim but that there was not enough room to write it. This note tormented mathematicians for centuries as they tried to solve what some have claimed to be the world's greatest mathematical problem.[9]

Fermat's last theorem stayed unsolved for more than 350 years, despite the attentions of some of the greatest minds in history. In recent years it was dramatically solved by a shy English mathematician, and the story of his work, told by a number of biographers, captures the drama, the intrigue, and the allure of mathematics that is unknown by many. Any child—or adult—wanting to be inspired by the values of determination and persistence, enthralled by the intrigue of puzzles and questions, and introduced to the sheer beauty of living mathematics should read Simon Singh's book *Fermat's Enigma*. Singh describes "one of the greatest stories in human thinking,"[10] providing important insights into the ways mathematicians work.

Many people had decided that there was no proof to be found of Fermat's theorem and that this great mathematical problem was unsolvable. Prizes were offered from different corners of the globe, and men and women devoted their lives to finding a proof, to no avail. Andrew Wiles, the mathematician who would write his name into history books, first encountered Fermat's theory as a ten-year-old boy while reading in his local library in his hometown of Cambridge. Wiles described how he felt when he read the problem: "It looked so simple, and yet all the great mathematicians in history could not solve it. Here was a problem that I, as a ten-year-old, could understand and I knew from that moment that I would never let it go, I had to solve it."[11] Years later, Wiles earned a PhD in mathematics from Cambridge and then took a position in Princeton's mathematics department. But it was only some years later that Wiles realized that he could devote his life to the problem that had intrigued him since childhood. As Wiles set about trying to prove Fermat's last theorem, he retired to his study and started reading journals and gathering new techniques. He began exploring and looking for patterns, working on small areas of mathematics, and then standing back to see if they could be illuminated by broader concepts. Wiles worked on a number of techniques over the next few years, exploring different methods for attacking the problem. Some seven years after starting the problem, Wiles emerged from his study one afternoon and announced to his wife that he had solved Fermat's last theorem.

The venue that Wiles chose to present his proof of the 350-year-old problem was a conference at the Isaac Newton Institute in Cambridge, England, in 1993. Some people had become intrigued about Wiles's work, and rumors had started to filter through that he was actually going to present a proof of Fermat's last theorem. By the time Wiles arrived, there were more than two hundred mathematicians crammed into the room, and some had sneaked in cameras to record the historic event. Others—who could not get in—peered through windows. Wiles needed three lectures to present his work and at the conclusion of the last the room erupted into great applause. Singh described the atmosphere of the rest of the conference as "euphoric" with the world's media flocking to the institute. Was it possible that this great and historical problem had finally been solved? Barry Mazur, a number theorist and algebraic geometer, reflected on the event, saying that "I've never seen such a glorious lecture, full of such wonderful ideas, with such dramatic tension, and what a buildup. There was only one possible punch line." Everyone who had witnessed the event thought that Fermat's last theorem was finally proved. Unfortunately, there was an error in Wiles's proof, which meant that he had to plunge himself back into the problem. In September 1994, after more months of work, Wiles knew that his proof was complete and correct. Using many different theories, making connections that had not previously been made, Wiles had constructed beautiful new mathematical methods and relationships. Ken Ribet, a Berkeley mathematician whose work had contributed to the proof, concluded that the landscape of mathematics had changed and mathematicians in related fields could work in ways that had never been possible before.

Long vs short problem?

why?

Wiles's fascinating story is told in detail by Simon Singh and others. But what do such accounts tell us that could be useful in improving children's education? One clear difference between the work of mathematicians and schoolchildren is that mathematicians work on long and complicated problems that involve combining many areas of mathematics. This stands in stark contrast to the *vs* short questions that fill the hours of math classes and that involve the repetition of isolated procedures. Long and complicated prob-

Persistence

⇒ lems are important to work on for many reasons—one being that they encourage persistence, a critical trait for young people to develop that will stand them in good stead in life and work. In interviews, mathematicians often speak of their enjoyment in working on difficult problems. Diane Maclagan, a professor at the University of Warwick, was asked, "What is the most difficult aspect of your life as a mathematician?" She replied, "Trying to prove theorems." The interviewer then asked what the most fun is. "Trying to prove theorems," she replied.[12] Working on long and complicated problems may not sound like fun, but mathematicians find such work enjoyable because they are often successful. It is hard for schoolchildren to enjoy a subject if they experience repeated failure, which of course is the reality for many young people in mathematics classrooms. But the reason that mathematicians are successful

is that they have learned something very important—and very learnable. They have learned to problem solve.

Problem solving is at the core of mathematicians' work, as well as the work of engineers and others, and it starts with the making of a guess. Imre Lakatos, mathematician and philosopher, describes mathematical work as "a process of 'conscious guessing' about relationships among quantities and shapes."[13] Those who have sat in traditional math classrooms are probably surprised to read that mathematicians highlight the role of guessing, as I doubt whether they have ever experienced any encouragement to guess in their math classes. When an official report in the UK was commissioned to examine the mathematics needed in the workplace, the investigator found that estimation was the most useful mathematical activity.[14] Yet when children who have experienced traditional math classes are asked to estimate, they are often completely flummoxed and try to work out exact answers, then round them off to look like an estimate. This is because they have not developed a good *feel* for numbers, which would allow them to estimate instead of calculate, and also because they have learned, wrongly, that mathematics is all about precision, not about making estimates or guesses. Yet both are at the heart of mathematical problem solving.

After making a guess, mathematicians engage in a zigzagging process of conjecturing, refining with counterexamples, and then proving. Such work is exploratory and creative, and many writers draw parallels between mathematical work and art or music. Robin Wilson, a British mathematician, proposes that mathematics and music "are both creative acts. When you are sitting with a bit of paper creating mathematics, it is very like sitting with a sheet of music paper creating music."[15] Devlin agrees, saying that "Mathematics is not about numbers, but about life. It is about the world in which we live. It is about ideas. And far from being dull and sterile, as it is so often portrayed, it is full of creativity."[16]

The exhilarating, creative pathways that mathematicians follow as they solve problems are often hidden in the end point of mathematical work, which just shows results. These pathways cannot be the exact same ones as schoolchildren experience, as children need to be taught the methods they need, as well as to use them in the solving of problems, but neither should school mathematics be so different as to be unrecognizable. As George Pólya, the eminent Hungarian mathematician, reflected,

> A teacher of mathematics has a great opportunity. If he fills his allotted time with drilling his students in routine operations he kills their interest, hampers their intellectual development, and misuses his opportunity. But if he challenges the curiosity of his students by setting them problems proportionate to their knowledge, and helps them to solve their problems with stimulating questions, he may give them a taste for, and some means of, independent thinking.[17]

Another interesting feature of the work of mathematicians is its collaboratory nature. Many people think of mathematicians as people who work in isolation, but this is far from the truth. Leone Burton, a British professor of mathematics education, interviewed seventy research mathematicians and found that they generally challenged the solitary stereotype of mathematical work, reporting that they preferred to collaborate in the production of ideas. More than half of the papers they submitted to Burton as representative of their work were written with colleagues. The mathematicians interviewed gave many reasons for collaboration, including the advantage of learning from one another's work, increasing the quality of ideas, and sharing the "euphoria" of problem solving. As Burton reflected, "They offered all the same reasons for collaborating on research that are to be found in the educational literature advocating collaborative work in classrooms."[18] Yet silent math classrooms continue to prevail across America.

Something else that we learn from various accounts of mathematicians' work is that an important part of real, living mathematics is the posing of problems. Viewers of *A Beautiful Mind* may remember John Nash (played by Russell Crowe) undergoing an emotional search to form a question that would be sufficiently interesting to be the focus of his work. People commonly think of mathematicians as solving problems, but as Peter Hilton, an algebraic topologist, has said: "Computation involves going from a question to an answer. Mathematics involves going from an answer to a question."[19] Such work requires creativity, original thinking, and ingenuity. All the mathematical methods and relationships that are now known and taught to schoolchildren started as questions, yet students do not see the questions. Instead, they are taught content that often appears as a long list of answers to questions that nobody has ever asked. Reuben Hersh puts it well: *Math started as questions*

> The mystery of how mathematics grows is in part caused by looking at mathematics as answers without questions. That mistake is made only by people who have had no contact with mathematical life. It's the questions that drive mathematics. Solving problems and making up new ones is the essence of mathematical life. If mathematics is conceived apart from mathematical life, of course it seems—dead.[20]

Bringing mathematics back to life for schoolchildren involves giving them a sense of living mathematics. When students are given opportunities to ask their own questions and to extend problems into new directions, they know mathematics is still alive, not something that has already been decided and just needs to be memorized. If teachers pose and extend problems of interest to students, they enjoy mathematics more, they feel more ownership of their work, and they ultimately learn more. English schoolchildren in math classes used to work on long problems that they could extend into directions that were of interest to them. For example, in one problem students were asked to design any type of building. This gave them the opportunity to consider interesting questions involving high-level mathematics, such as the best design for a fire station with a firefighter's pole. Teachers would submit the students' work to examination boards, and it was assessed as part of the students' final examination grades. When I asked English schoolchildren about their work on these problems, they reported that they were not only enjoyable and they learned a lot from them but also that their work made them "feel proud" and that they could not feel proud of their more regular textbook work.

Another important part of the work of mathematicians that enables successful problem solving is the use of a range of representations such as symbols, words, pictures, tables, and diagrams, all used with precision. The precision required in mathematics has become something of a hallmark for the subject and it is an aspect of mathematics that both attracts and repels. For some schoolchildren it is comforting to be working in an area where there are clear rules for ways of writing and communicating. But for others it is just too hard to separate the precision of mathematical language with the uninspiring drill-and-kill methods that they experience in their math classrooms. There is no reason that precision and drilled teaching methods need to go together, and the need for precision with terms and notation does not mean that mathematical work precludes open and creative exploration. On the contrary, it is the fact that mathematicians can rely on the precise use of language, symbols, and diagrams that allows them to freely explore the *ideas* that such communicative tools produce. Mathematicians do not play with the notations, diagrams, and words as a poet or artist might. Instead, they explore the relations and insights that are revealed by different arrangements of the notations. As Keith Devlin reflects:

> Mathematical notation no more *is* mathematics than musical notation is music. A page of sheet music represents a piece of music, but the notation and the music are not the same; the music itself happens when the notes on the page are sung or performed on a musical instrument. It is in its performance that the music comes alive; it exists not on the page but in our minds. The same is true for mathematics.[21]

Mathematics is a performance, a living act, a way of interpreting the world. Imagine music lessons in which students worked through hundreds of hours of sheet music, adjusting the notes on the page, receiving checks and crosses from the teachers, but never playing the music. Students would not continue with the subject because they would never experience what music *is*. Yet this is the situation that continues, seemingly unabated, in mathematics classes.

Those who use mathematics engage in mathematical *performances*. They use language in all its forms, in the subtle and precise ways that have been described, in order to do something with mathematics. Students should not just be memorizing past methods; they need to engage, do, act, perform, and problem solve, for if they don't *use* mathematics as they learn it, they will find it very difficult to do so in other situations, including examinations.

Maryam Mirzakhani is a mathematician at Stanford who was featured in newspapers across the world by winning the Fields Medal, the most prestigious prize in mathematics. When news of the award spread, the *Telegraph*, Britain's bestselling newspaper, asked me to write an article on Maryam's work. I was happy to do so,[22] and I told the newspaper that just weeks before I had chaired a PhD defense for one of Maryam's students in Stanford's math department. A PhD defense is the occasion when a student defends her PhD dissertation to a committee of scholars. That day I sat with the other members of the committee, all mathematicians, and watched as Maryam's student, a young woman, paced the room, showing visual representations and sharing conjectures, using mathematics creatively to connect different ideas. Many times in the defense she was asked a question to which she replied, "I don't know." This answer was perfectly reasonable and accepted by the mathematicians on the committee because she was exploring new territory to which no one had answers. This struck me as highly significant because schoolchildren everywhere would be shocked to see that high-level mathematics involves such creativity and uncertainty.

The erroneous thinking behind many school approaches is that students should spend years being drilled in a set of methods that they can use later. The mathematicians who oppose change are most concerned about the students who will enter graduate programs in mathematics. At that point students will encounter real mathematics and use the tools they have learned in school to work in new, interesting, and authentic ways. But by this time most students have given up on math. We cannot keep pursuing an educational model that leaves the best and the only real taste of the subject to the end, for the rare few who make it through the grueling years that precede it. If students were able to work for at least some of the time in the ways mathematicians do—posing problems, making conjectures using intuition, exploring with and refining ideas, and discussing ideas with others—then they would not only be given a sense of true mathematical work, which is an important goal in its own right,[23] they would also be given the opportunities to enjoy mathematics and learn it in the most productive way.[24, 25, 26]

2 / What's Going Wrong in Classrooms?

Identifying the Problems

The Night of Nonsense

Soon after I first became a Stanford professor, I was introduced to what I believe to be a damaging and very strange phenomenon. The phenomenon—which began in California, but has now taken hold across the United States—is known as the math wars,[1] a series of unproductive and heated exchanges between advocates of different mathematics approaches. Participants in the math wars have been known to engage in hunger strikes, secret meetings, and extended campaigns of abuse, all in the pursuit of their preferred method of teaching. One outcome of the math wars is that good teachers have been driven out of the teaching profession after years of bullying by extremists. Another result is that any questions or discussions about change in mathematics teaching have been suppressed and paths to improvement have been blocked.

Ironically, the issue that has gotten so many people hot under the collar is not even the most important one. The subject of the math wars is the *curricula* that teachers use in their classrooms—the published sets of textbooks that schools and districts choose to adopt. Of course, the books used in classrooms are important and we want our children to be working with high-quality materials that teach mathematics well, but the most important factor in school effectiveness, proved by study after study, is not the curriculum but the teacher.[2, 3, 4, 5] Good teachers can make mathematics exciting even with a dreary textbook. Conversely, bad teachers do not become good just because a book is written well, but the math wars have prevented people from focusing upon and supporting good teaching; instead, they have, quite deliberately, turned attention away from teachers and tried to *force* particular curricula upon them. Emily Moskam, whose classroom I described in the introduction, has now left teaching. A teacher who could inspire children to love and use mathematics, at the highest levels, like no other I have seen could not stand being forced to use textbooks that she knew were ineffective and that were damaging her students' learning. She could see no way forward and ultimately chose to leave the profession.

I experienced my first personal contact with the math wars on a cool November evening in California. I was thinking about including Emily Moskam's high school in my research study when I learned of a meeting taking place at her school to discuss its mathematics curriculum. Strangely, although the meeting was for parents to consider the school's math approach, no teachers or administrators had been invited. As parents walked into the hall they saw three women standing at the front, passing papers back and forth. All three were parents of freshmen in the school and one later confided to me that she had spent a year gathering data for that meeting.

The meeting had been arranged because the math department at Emily's school had abandoned the traditional books and methods of teaching that they had used for many years, with very little success, and started using an award-winning curriculum that engaged students in real—and impressively hard—problems to solve (IMP). Students responded well to the new curriculum. They reported enjoying math more and many more of them were taking high-level classes. The teachers told me they were the happiest they had ever been—they were attending professional development workshops on weekends as well as spending time discussing good teaching methods with one another, and their students were doing well. It was around this time that a group of extreme traditionalists heard about the school's new approach and started to plot its downfall. They needed parents at the school to be the public face of their attack, and the women who were running the meeting that night had stepped forward.

In that first momentous meeting the women bombarded the gathered parents with data that had been fed to them, telling them that if their children continued with the new math program, then they would not be eligible to go to college and that test scores would fall. They showed graphs that had been constructed to give the impression of falling student test scores. It is easy to manipulate educational data to give the illusion of a bad program, as it is always possible to find a set of students somewhere whose scores have declined—even if for a particular reason or for a very brief period of time—and then to produce graphs with inappropriate axes that make tiny differences appear huge or that generalize from ten students. To support their claim that students would be ineligible for college, the women had phoned a range of prestigious colleges and asked them a question to the effect of "Would you accept a student who had not taken any math in high school but had just talked about math?" Many of the colleges said no and the women started composing a list. These tactics may sound incredible, but the people involved felt justified in trying to promote their position, using any method that they could, as they believed that they were involved in a "great educational war"[6] and that any tactics, no matter how underhanded, are admissible when at war.

That night they also suggested something to the parents that greatly upset the teachers when they heard about it—they implied that the teachers were being paid by the program and they were supporting it for personal gain. The parents left the hall in a somber mood that night, some of them skeptical, many of them scared. I was amazed that such a meeting could take place, but that was just the beginning.

In the weeks that followed, many of the parents, sensibly, asked for proper data. They knew of other students who had taken this math approach and been very successful. They contacted some of the colleges that would, apparently, not accept students from the math program and they found out that the admissions officers had not been asked about the program, only a ridiculous question about talking instead of learning math. Stanford University was one of the universities being used as an example of a university that would not accept students who had studied with the new curriculum approach—despite having admitted many students who had studied with the curriculum the teachers were using. When Stanford heard that it was being used as an example of a university that would not admit such students, its admissions officers quickly wrote a letter saying that this was not true and that the admissions office did not discriminate against any high school math program. Unfortunately, the damage had already been done.

It was around this time that I was sitting in a local coffee shop with a group of my graduate students discussing the research we were planning to conduct at the school. A girl and her mother walked over to our table. The girl asked us: "Are you discussing that new math program at the school? It has ruined my life!" When we asked her why, she and her mother explained that she would no longer be eligible to go to college. We explained to them both that this was not true. They thanked us for the information, but they continued to look upset.

Sensing that they had not won over all of the parents, the three women tried another line of attack. They moved on to the students, trailing after them at break times, telling them they would be ineligible for college, and asking them to sign a petition to end the math program. The students became more and more confused and scared, so many of them signed the petitions. By the time the teachers were told anything about the secret campaign, it was all over. The women, supported and organized by the extreme traditionalists, had scared enough parents and convinced the school board that the teachers must go back to teaching traditionally. Now desks are in rows, teachers lecture, students silently copy methods and then practice with lots of examples, and the problem solving that students used to love is no longer in evidence. The teachers at the school were demoralized and defeated.

There were two types of books that were at the center of this. Both introduced students to the same mathematical methods and procedures, but the IMP books attempt to do so in a way that made the mathematics more meaningful for students. Consider, for example, the way that the IMP book and the traditional book (brought in by the parent activists) introduced algebraic variables to the students. The traditional book begins with a brief example of rental charges at an ocean shop and then has the following text:

The letter h stands for the hours shown in the table: 1, 2, 3, or 4. Also, h can stand for other hours not in the table. We call h a variable.

A **variable** is a symbol used to represent one or more numbers. The numbers are called **values of the variable**. An expression that contains a variable, such as the expression $4.50 \times h$, is called a **variable expression**. An expression that names a particular number, such as 4.50×4, is called a **numerical expression**, or **numeral**.

Another way to indicate multiplication is to use a raised dot, for example, $4.50 \cdot 4$. In algebra, products that contain a variable are usually written without the multiplication sign because it looks too much like the other x, which is often used as a variable.[7]

The traditional book gives two pages of explanations like these before presenting twenty-six "oral exercise" questions and forty-nine "written exercise" questions,[8] such as:

$$\text{Simplify } 9 + (18 - 2) \text{ and}$$
$$2 \cdot (b + 2)$$

In contrast, the IMP book introduced students to variables through a particular situation—that of the nineteenth-century settlers who traveled from Missouri to California to set up life on the West Coast. The students are told, "You will encounter some very important mathematical ideas—such as graphs, different use of variables, lines of best fit, and rate problems—as you travel across the continent."[9] They are then given various exercises that require them to represent the settlers' situations using mathematical tools such as graphs and variables. For example, the students are told about families and some general conventions, such as "anyone more than fourteen years old is considered an adult."[10] They are then introduced to variables through this question:

The Hickson household contains 3 people of different gen-
erations. The total age of the 3 family members is 90.

a) Find reasonable ages for the 3 Hicksons.
b) Find another reasonable set of ages for them.

One student in solving this problem wrote:

$$C + (C + 20) + (C + 40) = 90$$

What do you think C means here?
How do you think the student got 20 and 40?
What set of ages do you think the student came up with? Try
this question, it can be solved in many ways.[11]

In the IMP curriculum, students are *gradually* introduced to
the concept of a variable—one of the most important concepts
in the mathematics curriculum—before being asked to interpret
and use variables in representing a situation mathematically.
The students are also encouraged to *discuss* variables, exploring their meaning and when they are used, and generally to
raise any questions or problems they have. The traditional curriculum, by contrast, tells students what variables are on page 1
of the book and then leads students through seventy-five practice questions. In requiring that students consider *situations*
and that they discuss the *meaning* of concepts such as variables, the IMP curriculum focuses more on understanding and
less on practicing methods. In the traditional curriculum, students practice a lot more. And this is where the crux of the
disagreement lies, with one group of people believing that students need to spend a lot of time practicing, and the other
group believing that it is better to understand an idea than it is
to practice it by rote.

My aim in this book is not to promote a curriculum. I am
aware that traditional and nontraditional books can be taught

well or badly and that any book requires a knowledgeable and caring teacher. But we would not be facing our current crisis if the people who were concerned about the newer books had worked with the mathematicians to improve them, rather than declaring war.

The wars in California were instigated by organizations such as Mathematically Correct, which hosted a Web site explaining that they are involved in a "great educational war" to save traditional math teaching. The site, which included no contact names or people, was full of articles attacking any new mathematics approaches and gave instructions on ways to get rid of any reforms in schools. Behind the Web site was a group of activists who scoured the country, looking for schools that were using new books so that they could land in the area, with a wealth of resources, and organize parents to crush the reforms. One of the most active members of Mathematically Correct wrote several threatening messages to me, because my published research has shown that students need opportunities to learn actively. He wrote that I pose a great threat, as I am a professor from a top university and have data. He recommended that readers of his Web pages visit university education departments and "nuke 'em all, dammit."[12] Others have told me that I had "better not" talk about my research publicly in America. Such threats and attempts to suppress research evidence may seem incredible, but they are characteristic of the events that make up the math wars.[13, 14] For more detailed accounts of this phenomenon that continues to suppress progress in the education of our children, I recommend reading University of California, Berkeley, professor Alan Schoenfeld's article "The Math Wars" and Michigan State professor Suzanne Wilson's book *California Dreaming*, both very readable accounts of a truly unfortunate set of events.

How It All Started

In the 1980s there was widespread awareness that students were failing mathematics in shockingly high numbers, and a range of reforms were introduced into schools, prompted by the National Council of Teachers of Mathematics (NCTM), which issued a new set of curriculum standards in 1989. Math books were quickly rewritten by publishers and filled with bright colors and real-world contexts. Teachers were instructed to be facilitators rather then lecturers and to have children work in groups. The reforms were introduced relatively quickly and often without consultation with parents. Some teachers were not trained to work in the new ways and so found it difficult. Critics claimed that mathematics was being threatened, that students were no longer learning standard methods, and that they were wasting time in groups chatting with friends instead of working. But rather than opening a dialogue between the different people who cared about mathematics teaching, battle lines were drawn and certain organizations such as Mathematically Correct declared war. The same groups are now fighting the introduction of Common Core Mathematics.

Learning without Thought

Those involved in the math wars think of different versions of mathematics teaching as either traditional or reform, and debates revolve around these two imaginary poles. In my research I have found that such categories do not actually mean much and that both camps include many types of teachers, teaching, and methods, some of which are highly effective and some not. Certain teachers might be described as traditional because they lecture and they have students work individually, but they also ask students great questions, engage them in interesting mathematical inquiries, and give students opportunities to solve problems, not just rehearse standard methods. Such teachers are wonderful and I wish there were many more of them. The type of traditional teaching that concerns me greatly and that I have identified from decades of research as highly ineffective is a version that encourages *passive learning*. In many mathematics classrooms across America the same ritual unfolds: teachers stand at the front of class demonstrating methods for twenty to thirty minutes of class time each day while students copy the methods down in their books, then students work through sets of near-identical questions, practicing the methods. Students in such classrooms quickly learn that *thought* is not required in math class and that the way to be successful is to watch the teachers carefully and copy what they do. In interviews with hundreds of students from such classes, I have asked them what it takes to be successful in math class, and they almost always give the exact same answer: *pay careful attention*. As one of the girls I interviewed told me, "In math you have to remember; in other subjects you can think about it."

Students taught through passive approaches follow and memorize methods instead of learning to inquire, ask questions, and solve problems. I have interviewed hundreds of students taught in such ways and they usually reflect on their experiences saying such things as "I'm just not interested in, just, you give me a formula, I'm supposed to memorize the answer, apply it, and that's it," and "You have to be willing to accept that sometimes things don't look like—they don't seem that you should do them. Like they have a point. But you have to accept them." Students who are taught using passive approaches do not engage in sense making, reasoning, or thought (acts that are critical to an effective use of mathematics), and they do not view themselves as active problem solvers. This passive approach, which characterizes math teaching in America, is widespread and ineffective.

When students try to memorize hundreds of methods, as

students do in classes that use a passive approach, they find it extremely hard to use the methods in any new situations, often resulting in failure on exams as well as in life. The secret that good mathematics users know is that only a few methods need to be memorized and that most mathematics problems can be tackled through the understanding of mathematical concepts and active problem solving. I am now working with the PISA (Programme for International Student Assessment) team, based in the OECD (Organisation for Economic Co-operation and Development) in Paris. The PISA team not only collects data on students' math achievements, but it also collects data on the strategies students use in mathematics and relate these to performance.[15] One of the strategies the PISA team has identified is "memorization." Some students think their role in math classrooms is to memorize all the steps and methods. Other students think their role is to connect ideas. These different strategies link, unsurprisingly, to achievement, and the students who memorize are the lowest achieving in the world. The highest-achieving students are the ones who think about the big ideas in mathematics. Unfortunately, in most US classrooms mathematics is presented as a set of procedures, leading students to think that their role in learning mathematics is to memorize. This PISA result gives us an interesting insight into the low performance of students in the United States, relative to the other countries tested.

I have spent my research career conducting unusual studies of learning. They are unusual because, instead of dropping in on students to see what they are doing in math classes, I have followed students through years of middle and high school, performing *longitudinal* studies. I have spent thousands of hours, with teams of graduate students, collecting data on students learning mathematics in different ways. This has included watching hundreds of hours of math classes, interviewing students about their experiences, giving them questionnaires on mathematical beliefs, and performing assessments to probe students' understanding. These studies have revealed that many math classrooms leave students cold, disinterested, or traumatized. In hundreds of interviews with students who have experienced passive approaches, they have told me that thought is not required, or even *allowed,* in math class. Children emerge from passive approaches believing that they only have to be obedient and memorize what the teacher tells them to do. They learn that they must simply memorize methods even when methods do not make sense. It is ironic that math—a subject that should be all about inquiring, thinking, and reasoning—is one that students have come to believe requires *no thought*.

In 1982, before teaching reforms were introduced in classrooms, students were asked in a national assessment to estimate the answer to

$$\frac{12}{13} + \frac{7}{8}$$

and they were given a choice of answers:

1, 2, 19, or 21.

Both numbers, 12/13 and 7/8, are close to 1, so an estimate tells us that their sum is close to 2. In the national assessment only 24 percent of thirteen-year-olds and a stunning 37 percent of seventeen-year-olds answered the question correctly. Most students chose the nonsensical answers of 19 or 21.[16] The seventeen-year-olds did not appear to make sense of the question and estimate, probably because they were trying to follow a rule and they made mistakes in the enactment of the rule.

The fact that students are drilled in methods and rules that

do not make sense to them is not just a problem for their understanding of mathematics. Such an approach leaves students
frustrated, because most of them want to understand what they
are learning. Students want to know how different mathematical methods fit together and why they work. This is especially
true for girls and women, as I shall explain in chapter 6. The
following response from Kate, a girl taking calculus in a traditional class, resembles those that I have received from many
young people I have interviewed:

> We knew how to do it. But we didn't know why we were
> doing it and we didn't know how we got around to doing
> it. Especially with limits, we knew what the answer was,
> but we didn't know why or how we went around doing
> it. We just plugged into it. And I think that's what I really
> struggled with—I can get the answer, I just don't under-
> stand why.

Young people are naturally curious and their inclination—at
least before they experience traditional teaching—is to make
sense of things and to understand them. Most American math
classes rid students of this worthy inclination. Kate was at least
fortunate to still be asking "Why?" even though she, like others,
was not given opportunities to understand why the methods
worked. Children begin school as natural problem solvers and
many studies have shown that students are better at solving
problems *before* they attend math classes.[17, 18] They think and
reason their way through problems, using methods in creative
ways, but after a few hundred hours of passive math learning
students have their problem-solving abilities drained out of them.
They think that they need to remember the hundreds of rules
they have practiced and they abandon their common sense in
order to follow the rules.

Consider for a moment this mathematics problem:

A woman is on a diet and goes into a shop to buy some turkey slices. She is given 3 slices that together weigh ⅓ of a pound, but her diet says that she is allowed to eat only ¼ of a pound. How much of the 3 slices she bought can she eat while staying true to her diet?

This is an interesting problem and I urge readers to try it before moving on. It was posed by Ruth Parker, a wonderful teacher of teachers who has spent many years working with parents to help them understand the benefits of inquiry approaches. In one of her public sessions with children and parents, she posed this problem and asked people to solve it. Her purpose in doing so was to see what kind of solutions people offered and how these compared to their school experiences. Many of the adults who had experienced passive approaches were unable to solve the problem because they could not apply a rule they had learned. Some of them tried $1/4 \times 1/3$, as they knew that something should be multiplied, but they recognized that their answer of $1/12$ was probably incorrect. Some tried $1/4 \times 3$, but their answer of $3/4$ of a pound also did not make sense. To use a rule they needed to set up the following equation:

$$3 \text{ slices} = ⅓$$
$$x \text{ slices} = ¼$$

Once Ruth told them this, the people who had remembered rules and methods were able to do the rest—to cross multiply and say that

$$⅓ \, x = ¾$$
$$\text{so } x = ¾ \text{ slices}$$

But as she pointed out, the most important part of the mathematics that is needed is to be able to set up the equation. This is something that children get very little experience in—they either use the same equation over and over again in a math lesson and so do not focus on how to set one up, or they are given equations that are already set up for them and they rehearse how to solve them, over and over again.

But look at some of the wonderful solutions offered by young children who had not yet been subjected to rule-bound passive approaches at school:

One fourth grader said,

If 3 slices is ⅓ of a pound, then 9 slices is a pound. I can eat ¼ of a pound, and ¼ of 9 slices is 9/4 slices.

Another solved the problem visually by representing a pound:

and then a quarter of a pound:

These elegant solutions involve the sorts of methods that are suppressed by passive, rule-bound mathematics approaches that teach only one way to solve problems and discourage all others. We can only speculate as to whether these same young people would be able to think of solutions such as these after future years of passive mathematics approaches. The fact that many students learn to suppress their thoughts, ideas, and problem-solving abilities in math classes is one of the most serious problems in American math education.

Learning without Talking

Another major problem with passive approaches to mathematics is that students work in silence. This may, to some, seem to be the optimum learning condition, but in fact this is far from the truth. I have visited hundreds of classrooms in which students sit in rows at individual desks, silently watching the teacher work on math and silently copying the methods. But this approach is flawed for a number of reasons. One problem is that students often need to talk through methods to know whether they really understand them. Methods can *seem* to make sense when people hear them, but explaining them to someone else is the best way to know whether they are really understood.

When two famous mathematicians from very different circumstances reflected on the conditions that allowed them to succeed in mathematics, I was struck by the similarity in their statements. Sarah Flannery is a young Irish woman who won the European Young Scientist of the Year award for the development of a "breathtaking" mathematical algorithm. In her autobiography she writes about the different conditions that promoted her learning, including the "simple math puzzles" that she worked on as a child, which I shall talk more about in chapter 8. Flannery writes: "The first thing I realized about learning mathematics was that there is a hell of a difference between, on the one hand, *listening* to math being talked about by somebody else and thinking that you are understanding, and, on the other, thinking about math and understanding it yourself and *talking* about it to someone else."[19] Reuben Hersh, an American mathematician, wrote the book *What Is Mathematics, Really?* In it he also talks about the source of his mathematical understanding, saying that "Mathematics is learned by computing, by solving problems, and by *conversing,* more than by reading and *listening*."[20]

Both of these successful mathematicians highlight the role of talking over listening, yet listening is the signature of the passive mathematics approaches that are the norm for American students. The first faulty learning condition that Flannery describes ("listening to math being talked about by someone else") is the quintessence of passive math approaches. The second condition that enabled her to understand ("thinking about math" and "talking about it to someone else") is what students should be doing in classrooms and homes and is essential to the active approach I will set out in chapter 3. When students listen to someone laying out mathematical facts (a passive act that does not necessarily involve intellectual engagement), they usually think that it makes sense, but such thoughts are very different from understanding, as the following true story illustrates.

Soon after Emily's school converted back to traditional classes, I visited an algebra lesson there. The teacher was an "old-school" math teacher who had been hired to teach the traditional approach. He strode up and down at the front of the room, filling the board with mathematical methods that he explained to students. He joked occasionally and peppered his sentences with phrases such as "this is easy" and "just do this quickly." The students liked him because of his jokes, cheery outlook, and clear explanations, and they would watch and listen carefully and then practice the methods in their books. The students had all been involved in the very public debate at the school between those supporting a traditional math approach and those supporting the newer curriculum materials. One day when I was visiting the algebra class that the students knew as "traditional math," I stopped and knelt by the side of one boy and asked him how he was getting on. He replied enthusiastically, "Great. I love traditional math. The teacher tells it to you and you get it." I was about to go to another desk when the teacher came around handing tests back. The boy's face fell as he saw a large F circled in red on the front of his test. He stared at the F, looked through his test, and turned back to me, saying, "Of course, that's what I hate about traditional math—you think you've got it when you haven't!" This afterthought, given with a wry smile, was amusing, but it also communicated something very important about the limitations of the approach that he was experiencing. Students do think that they "get it" when methods are shown to them on the board and they repeat them lots of times, but there is a huge difference between seeing something that appears to make sense and understanding it well enough to use it a few weeks or days later or in different situations. To know whether students are understanding methods as opposed to just thinking that everything makes sense, they need to be solving complex problems—not just repeating procedures with different numbers— and they need to be talking through and explaining different methods.

Another problem with the silent approach is that it gives students the wrong idea about mathematics. One of the most important parts of being mathematical is an action called *reasoning*. This involves explaining *why* something makes sense and how the different parts of a mathematical solution lead from one to another. Students who learn to reason and to *justify* their solutions are also learning that mathematics is about making sense. Reasoning is central to the discipline of mathematics. Scientists work to prove or disprove theories by finding new cases, but mathematicians prove their work through reasoning. Mathematicians make logically connected statements and reason their way to a proof. When parents ask me why their children should "waste time" in class "explaining their work" when they "know the answers," I tell them that they need to explain their work because this is the most mathematical of acts. If students are not reasoning, then they are not thinking and working mathematically. Whenever students offer a solution to a math problem, they should know why the solution is appropriate, and they should draw from mathematical rules and principles when they justify the solution rather than just saying that a textbook or a teacher told them it was right. Reasoning and justifying are both critical acts, and it is very difficult to engage in them without talking. If students are to learn that being mathematical involves making sense of their work and being able to explain it to someone else, justifying the different moves, then they need to talk to one another and to their teacher.

Another reason that talking is so critical in mathematics classrooms is that when students discuss mathematics, they come to know that the subject is more than a collection of rules and methods set out in books—they realize that mathematics is a subject that they can have their own ideas about, a subject that can invoke different perspectives and methods, and a subject that is connected through organizing concepts and themes. This is important for all learners but perhaps none more so than adolescents. If young people are asked to work in silence and they are not asked to offer their own ideas and perspectives, they often feel disempowered and disenfranchised, ultimately choosing to leave mathematics even when they have performed well.[21] When students are asked to give their ideas on mathematical problems, they feel that they are using their intellect and that they have responsibility for the direction of their work, which is extremely important for young people.

Mathematical discussions are also an excellent resource for student understanding. When students explain and justify their work to each other, they get to hear each other's explanations, and there are times when students are much more able to understand an explanation from another student than from a teacher. The students who are talking are able to gain deeper understanding through explaining their work, and the ones listening are given greater access to understanding. One of the reasons for this is that when we verbalize mathematical thoughts, we need to reconstruct them in our minds, and then when others react to them, we reconstruct them again. This act of reconstruction deepens understanding.[22] When we work on mathematics in solitude, there is only one opportunity to understand the mathematics. Of course, discussions need to be organized well. I will explain in later chapters how successful discussions are managed.

The bottom line on talking is that it is critical to math learning and to giving students the depth of understanding they need. This does not mean that students should be talking all the time or that just any form of talking is helpful. Math teachers need to organize productive mathematical discussions, and they should give children some time to discuss math and some time to work alone. But the distorted version of mathematics that is conveyed in silent math classrooms is one that makes mathematics inaccessible and extremely boring to most children.

Learning without Reality

A problem with both old and newer mathematics approaches that has emerged from my research is the ridiculous problems that are used in mathematics classrooms. Just like stepping through the wardrobe door and entering Narnia, in math classrooms trains travel toward each other on the same tracks and people paint houses at identical speeds all day long. Water fills tubs at the same rate each minute, and people run around tracks at the same distance from the edge. To do well in math class, children know that they have to suspend reality and accept the ridiculous problems they are given. They know that if they think about the problems and use what they understand from life, then they will fail. Over time, schoolchildren realize that when you enter Mathland you leave your common sense at the door.

Contexts started to become more common in math problems in the 1970s and '80s. Up until that point most mathematics had been taught through abstract questions with no reference to the world. The abstractness of mathematics is synonymous for many people with a cold, detached, remote body of knowledge. Some believed that this image may be broken down through the use of contexts, and so mathematics questions were placed into contexts with the best of intentions. But instead of giving students realistic situations that they could analyze, textbook authors began to fill books with make-believe contexts—contexts that students were meant to believe but for which they should not use any of their real-world knowledge. Students are frequently asked to work on questions involving, for example, the price of food and clothes, the distribution of pizza, the numbers of people who can fit into an elevator, and the speeds of trains as they rush toward each other, but they are not meant to use any of their actual knowledge of clothing prices, people, or trains. Indeed, if they do engage in the questions and use their real-world knowledge, they will fail. Students come to know this about math class. They know that they are entering a realm in which common sense and real-world knowledge are not needed.

Here are the sorts of examples that fill math books:

> Joe can do a job in 6 hours and Charlie can do the same job in 5 hours. What part of the job can they finish by working together for 2 hours?

> A restaurant charges $2.50 for ⅛ of a quiche. How much does a whole quiche cost?

> A pizza is divided into fifths for 5 friends at a party. Three of the friends eat their slices, but then 4 more friends arrive. What fractions should the remaining 2 slices be divided into?

Everybody knows that people work together at a different pace than when they work alone, that food sold in bulk such as a whole quiche is usually sold at a different rate than individual slices, and that if extra people turn up at a party more pizza is ordered or people go without slices—but none of this matters in Mathland. One long-term effect of working on make-believe contexts is that such problems contribute to the mystery and otherworldliness of Mathland, which curtails people's interest in the subject. The other effect is that students learn to ignore contexts and work only with the numbers, a strategy that would not apply to any real-world or professional situation. An illustration of this phenomenon is given by this famous question asked in a national assessment of students:

An army bus holds 36 soldiers. If 1,128 soldier are being bussed to their training site, how many buses are needed?

The most frequent response from students was 31 remainder 12, a nonsensical answer when dealing with the number of buses needed.[23] Of course, the test writers wanted the answer of 32, and the "31, remainder 12" response is often used as evidence that American students do not know how to interpret situations. But it may equally be given as evidence that they have been trained in Mathland, where such responses are sensible.

My argument against pseudocontexts does not mean that contexts should not be used in mathematics examples; they can be extremely powerful. But they should only be used when they are realistic and when the contexts offer something to the students, such as increasing their interest or modeling a mathematical concept. A realistic use of context is one where students are given real situations that need mathematical analysis, for which they do need to consider (rather than ignore) the variables. For example, students could be asked to use mathematics to predict population growth. This would involve interpreting newspaper data on the US population, investigating the amount of growth over recent years, determining rates of change, building linear models ($y = mx + b$), and using these to predict population growth into the future. Such questions are excellent ways to interest students, motivate them, and give them practice in using mathematics to solve problems. Contexts may also be used to give a visual representation, helping to convey meaning. It does not hurt to suggest that a circle is a pizza that needs dividing into fractions, but it does hurt when students are invited into the world of parties and friends while at the same time being required to ignore everything they know about parties and friends.

There are also many wonderful mathematics problems with

no context or barely any context at all that can engage students. The famous four-color problem, which intrigued mathematicians for centuries, is a good example of a gripping, abstract math problem. The problem came about in 1852 when Francis Guthrie was trying to color a map of the counties of England. He did not want to color any adjacent counties with the same color and noticed that he needed only four colors to cover the map. Mathematicians then set out to prove that only four colors would be needed in any map or any set of touching shapes. It took more than a century to prove this, and some still question the proof.

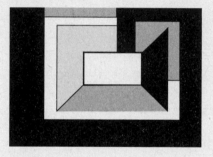

This is a great problem that can be given to students to investigate. They can work with a map of touching countries, such as Europe, or draw their own shapes. For example,

Can you color this map using only 4 colors, with no 2 adjacent "countries" colored the same?

Other examples of problems I have used in this book, like the chessboard problem in the introduction, use contexts sensibly and responsibly. The contexts give meaning to the problems and provide realistic constraints—students do not have to partly believe them and partly ignore them.

The artificiality of mathematics contexts may seem to be a small concern, but the long-term impact of such approaches can be devastating for students of mathematics. Hilary Rose, a sociologist and newspaper columnist, illustrates this point well. She recalls that as a young child she had been something of "a mathematical wunderkind"[24] and loved to explore patterns, numbers, and shapes. She describes how her sense of mathematical magic ended when real-world problems were used in school. At first she enthusiastically engaged with the problems, drawing upon her knowledge of the situations described, but then found that such engagement was not allowed:

> Thinking about it, it was those so called practical problems that irritated me the most. It was obvious to me that many of the questions simply indicated that the questioner did not know enough about the craft skills involved in real-world solutions. Lawn rollers being pulled up slopes, wallpapering rooms by calculating square feet and inches: these were tedious and as far as that highly practical child could see, stupid . . . I know that the price I paid was to lose my sense of confidence that school maths and everyday maths were part of one world.[25]

[Some of the studies I refer to in this book took place in the UK and Australia, where *mathematics* is shortened to *maths* as opposed to the American abbreviation *math*.]

If the ridiculous contexts were taken out of math books across America—be they the traditional or newer books—they would probably fall in size by more than half. The elimination of ridiculous contexts would be good for many reasons. Most important, students would realize that they are learning an important subject that helps make sense of the world, rather than a subject that is all about mystification and non-sense.

As the world changes and technology becomes more and more pervasive in our jobs and lives, it is impossible to know exactly which mathematical methods will be most helpful in the future. This is why it is so important that schools develop flexible thinkers who can draw from a variety of mathematical principles in solving problems. The only way to create flexible mathematical thinkers is to give children experience in working in these ways, both in school and at home. In the next chapter I will consider two very different school approaches that were extremely effective in achieving this goal.

ra2studio

3 / A Vision for a Better Future

Effective Classroom Approaches

Imagine a time when children were eager to go to math lessons at school, excited to learn new mathematical ideas, and able to use math to solve problems outside of the classroom. Adults would feel comfortable with math, be happy to be given mathematical problems at work, and refrain from saying at parties, "I am terrible at math." America would have the number and range of people good at math to fill the various jobs needing mathematical and scientific understanding that our technological age requires. All of this might sound far-fetched given the number of mathematically damaged and phobic people in the population and the scores of schoolchildren who dread math lessons. But a very different mathematical reality is achievable, and teachers and parents can bring about this reality. In the pages that follow I will describe two highly successful approaches that offered students an experience of real mathematical work. They are middle and high school level, but the principles of the two approaches apply to all levels. In the following pages we will learn about students from a wide range of backgrounds who came to love math, to achieve at high levels, and to view mathematics as an important part of their future.

The Communicative Approach

Photo courtesy of Jo Boaler

Railside High School is an urban school in California and lessons are frequently interrupted by the sound of speeding trains. As with many urban schools, the buildings look as though they are in need of repair, but Railside is not like other urban schools. Calculus classes are often badly attended or nonexistent in many schools, but at Railside they are packed with eager and successful students. When I have taken visitors to the school and we have stepped inside the math classrooms, they have been amazed to see all the students hard at work, engaged, and excited about math. I first visited Railside because I had learned that the teachers collaborated and planned teaching ideas together, and I was interested to see their lessons. I saw enough in that visit to invite the school to be part of a new Stanford research project investigating the effectiveness of different mathematics approaches. Some four years later, after we had followed seven hundred students through three high schools, observing, interviewing, and assessing them, we knew that Railside's approach was both highly successful and highly unusual.

The mathematics teachers at Railside used to teach using traditional methods, but they were unhappy with their students' high failure rates and lack of interest in math, so they worked together to design a new approach. Teachers met over several summers to devise a new algebra curriculum and later to improve all the courses they offered. They also detracked classes and made algebra the first course that *all* students would take on entering high school—not just higher achieving students. In most algebra classes students work through questions designed to give practice in mathematical techniques such as factoring polynomials or solving inequalities. At Railside the students learned the same methods, but the curriculum was organized around bigger mathematical ideas, with unifying themes such as "What is a linear function?" The focus of the Railside approach was "multiple representations," which is why I have described it as communicative—the students learned about the different ways that mathematics could be communicated through words, diagrams, tables, symbols, objects, and graphs. As they worked, the students would frequently be asked to explain work to each other, moving between different representations and communicative forms. When we interviewed students and asked them what they thought math was, they did not tell us that it was a set of rules, as most students do. Instead, they told us that math was a form of communication, or a language. As one young man explained: "Math is kind of like a language, because it has got a whole bunch of different meanings to it, and I think it is communicating. When you know the solution to a problem, I mean that is kind of like communicating with your friends."

In one of the lessons I observed, students were learning about functions. The students had been given what the teachers referred to as "pile patterns." Different students had been given different patterns to work with. Pedro was given the following pattern, which includes the first three cases:

Case 1   Case 2   Case 3

The aim of the activity was for students to work out how the pattern was growing (you could try this too) and to represent this as an algebraic rule, a t-table, a graph, and a generic pattern. They also needed to show the hundredth case in the sequence, having been given the first three cases.

Pedro started by working out the numbers that went with the first 3 cases, and he put these in his t-table:

Case Number	Number of Tiles
1	10
2	13
3	16

He noted at this point that the pattern was "growing" by 3 each time. Next he tried to *see* how the pattern was growing in his shapes, and after a few minutes he saw it! He could see that each of the 3 sections grew by 1 each time. He represented the first two cases in this way:

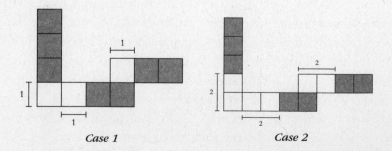

Case 1   Case 2

He could see that there were seven tiles that always stayed the same and were present in the same positions (this was the way he visualized the pattern growth, but there are other ways of visualizing it). In addition to the "constant" seven, there were lines of tiles that grew with every case number. So, for example, if we just look at the vertical line of tiles:

Case 1   *Case 2*

we see that in case 1, there is 1 at the bottom + 3. In case 2, there are 2 + 3. In case 3 there would be 3 + 3; and in case 4, there would be 4 + 3, and so on. The 3 is a constant, but there is one more added to the lower section of tiles each time. We can also see that the growing section is the same size as the case number each time. When the case is 1, the total number of tiles is 1 + 3; when the case is 2, the total is 2 + 3. We can assume from this that when it is the hundredth case, there will be 100 + 3 tiles. This sort of work—considering, visualizing, and describing patterns—is at the heart of mathematics and its applications.

Pedro represented his whole pattern in the following way:

where x stood for the case number. By adding together the three sections, he could now represent the whole function as $3x + 7$.

At this point I should explain something about algebra for those who find this example totally bewildering. When a friend of mine read about this pattern, she was utterly lost and I realized that her confusion came from the way she had learned algebra in her traditional math classes. She looked at the pattern with me, saw that the student had represented it as $3x + 7$, and asked me, "So what is x?" I said that x was the case number, so in the first case x is 1, in the second x is 2, and so on. This completely confused her, which I realized was because to her x was always meant to be a *single* number. She had spent so many years of math classes "solving for x"—rearranging equations to find out what number x was—that she, like millions of schoolchildren, had missed the most important point about algebra—that x is used to represent a *variable*. The reason that algebra is used so pervasively by mathematicians, scientists, medics, computer programmers, and many other professionals is because patterns—that grow and change—are central to their work and to the world, and algebra is a key method in describing and representing them. My friend could see that the pattern had a different number of tiles each time but was just not used to using algebra to represent a *changing* quantity. But the task in this problem—to find a way of visualizing and representing the pattern, using algebra to describe the changing parts of the pattern—is extremely important algebraic work. The way that most people learn algebra hides the meaning of algebra, it stops them from using it appropriately, and it hinders their ability to see the usefulness of algebra as a problem-solving tool in mathematics and science.

Pedro was pleased with his work and he decided to check his algebraic expression with his t-table. Satisfied that $3x + 7$ worked, he set about plotting his values on a graph. I left the group as he was eagerly reaching for graph paper and colored pencils. The next day in class I checked in with him again. He was sitting with three other boys and they were designing a poster to show their four functions. Their four desks were pushed together and covered by a large poster that was divided into four sections. From a distance the poster looked like a piece of mathematical artwork with color-coded diagrams, arrows connecting different representations to each other, and large algebraic symbols.

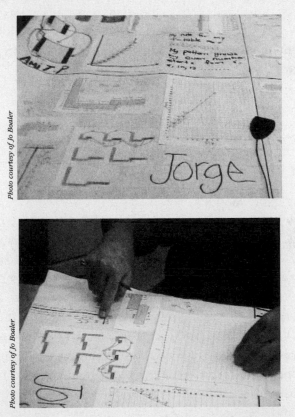

Photo courtesy of Jo Boaler

Photo courtesy of Jo Boaler

After a while the teacher came over and looked at the boys' work, talking with them about their diagrams, graphs, and algebraic expressions, and probing their thinking to make sure they understood the algebraic relationships. He asked Pedro where the 7 (from $3x + 7$) was represented on his graph. He showed the teacher and then decided to show the +7 in the same color on his tile patterns, on his graph, and in his algebraic expression. The communication of key features of functions using color coding was something all students were taught in the Railside approach, to give meaning to the different representations. This helped the students learn something important—that the algebraic expression shows something tangible and that the relationships within the expression can also be seen in the tables, graphs, and diagrams.

Juan, sitting at the same table, had been given a more complicated, nonlinear pattern that he had color coded in the following way:

See if you can work out how the pattern is growing and the algebraic expression that represents it!

As well as producing posters that showed linear and non-linear patterns, the students were asked to find and connect patterns, within their own pile patterns, and across all four teammates' patterns, and to show the patterns using technical writing tools. Because some of the pile patterns were nonlinear, this was a complicated task for ninth graders, and it provoked much discussion, consternation, and learning! One aim of the lesson was to teach students to look for patterns within and among representations and to begin to understand generalization.

The tasks at Railside were designed to fit within the separate subject areas of algebra and geometry, in line with the tradition of content separation in American high schools. Still, the problems the students worked on were open enough to be thought about in different ways, and they often required that students represent their thinking using different mathematical representations, emphasizing the connections between algebra and geometry.

The Railside classrooms were all organized in groups, and students helped each other as they worked. The Railside teachers paid a lot of attention to the ways the groups worked together, and they taught students to respect the contributions of other students, regardless of their prior achievement or their status with other students. One unfortunate but common side effect of some classroom approaches is that students develop beliefs about the inferiority or superiority of different students. In the other classes we studied, classes that were taught traditionally, students talked about other students as smart and dumb, quick and slow. At Railside the students did not talk in these ways. This did not mean that they thought all students were the same, but they came to appreciate the diversity of the class and the various attributes that different students offered. As Zane described to me: "Everybody in there is at a different level. But what makes the class good is that everybody's at different levels, so everybody's constantly teaching each other and helping each other out." The teachers at Railside followed an approach called complex instruction,[1] a method designed to make group work more effective and to promote equity in classrooms. They emphasized that all children were smart and had strengths in different areas and that everyone had something important to offer when working on math.

As part of our research project we compared the learning of the Railside students to that of similar-size groups of students in two other high schools who were learning mathematics through a more typical, traditional approach. In the traditional classes the students sat in rows at individual desks, they did not discuss mathematics, they did not represent algebraic relationships in different ways, and they generally did not work on problems that were applied or visual. Instead, the students watched the teacher demonstrate procedures at the start of lessons and then worked through textbooks filled with short, procedural questions. The two schools using the traditional approach were more suburban, and students started the schools with higher mathematics achievement levels than the students at Railside. But by the end of the first year of our research study, the Railside students were achieving at the same levels as the suburban students on tests of algebra; by the end of the second year, the Railside students were outperforming the other students on algebra and geometry tests.

In addition to the high achievement at Railside, the students learned to enjoy math. In surveys administered at various times during the four years of the study, the students at Railside were always significantly more positive and more interested in mathematics than the students from the other classes. By their senior year, a staggering 41 percent of the Railside students were in advanced classes of precalculus and calculus, compared with only 23 percent of students from the traditional classes. Further, at the end of the study, when we interviewed 105 students (mainly seniors) about their future plans, almost all of the students from the traditional classes said that they had decided not to pursue mathematics as a subject—even when they had been successful. Only 5 percent of students from the traditional classes planned a future in mathematics compared with 39 percent of the Railside students.

There were many reasons for the success of the Railside students. Importantly, they were given opportunities to work on interesting problems that required them to think (not just to reproduce methods), and they were required to discuss mathematics with one another, increasing their interest and enjoyment. But there was another important aspect of the school's approach that is much more rare—the teachers enacted an expanded conception of mathematics and "smartness." The teachers at Railside knew that being good at mathematics involves many different ways of working, as mathematicians' accounts tell us. It involves asking questions, drawing pictures and graphs, rephrasing problems, justifying methods, and representing ideas, in addition to calculating with procedures. Instead of just rewarding the correct use of procedures, the teachers encouraged and rewarded all of these different ways of being mathematical. In interviews with students from both the traditional and the Railside classes, we asked students what it took to be successful in math class. Students from the traditional classes were unanimous: they all said that it involved paying careful attention—watching what the teacher did and then doing the same. When we asked students from the Railside classes, they talked of many different activities such as asking good questions, rephrasing problems, explaining ideas, being logical, justifying methods, representing ideas, and bringing a different perspective to a problem. Put simply, because there were many more ways to be successful at Railside, many more students *were* successful.

Janet, one of the freshmen, described to me the way that Railside was different from her middle school experience: "Back in middle school the only thing you worked on was your math skills. But here you work socially and you also try to learn to help people and get help. Like you improve on your social skills, math skills, and logic skills." Another student, Jasmine, also talked about the variety in Railside's approach, saying that "with math you have to interact with everybody and talk to them and answer their questions. You can't be just like 'Oh, here's the book. Look at the numbers and figure it out.'" We asked Jasmine why math was like that. She answered: "It's not just one way to do it. . . . It's more interpretive. It's not just one answer. There's more than one way to get it. And then it's like 'Why does it work?'" The students highlighted the different ways that mathematics problems could be solved and the important role played by mathematical justification and reasoning. The students at Railside recognized that helping, interpreting, and justifying were critical and valued in their math classes.

In addition, the teachers at Railside were very careful about identifying and talking to students about all the ways they were smart. The teachers knew that students—and adults—are often severely hampered in their mathematical work by thinking they are not smart enough. They also knew that all students could contribute a great deal to mathematics, so they took it upon themselves to identify and encourage students' strengths. This paid off. Any visitor to the school would have been impressed by the motivated and eager students who believed in themselves and who knew they could be successful in mathematics.

The Project-Based Approach

Phoenix Park School

The day that I walked into Phoenix Park School, in a low-income area of England, I didn't know what to expect. I had invited the maths department at the school to be a part of my research project. I knew that the department used a project-based approach, but I did not know much more than that. I made my way across the playground and into the school buildings on that first morning with some trepidation. A group of students congregated outside their maths classroom at break time and I asked them what I should expect from the lesson I was about to see. "Chaos," said one of the students. "Freedom," said another. Their descriptions were curious, and they made me more excited to see the lesson. Some three years later, after moving with the students through school, observing hundreds of lessons, and researching the students' learning, I knew exactly what they meant.

This was to be my first longitudinal research project on different ways of learning mathematics and, as with the Railside study, I watched hundreds of hours of lessons, interviewed and gave surveys to students, and performed various assessments. I chose to follow an entire cohort of students in each of two schools, from when they were thirteen to when they were sixteen years of age. One of the schools, Phoenix Park, used a project-based approach, the other, Amber Hill, used the more typical traditional approach. The two schools were chosen because of their different approaches, but also because their student intakes were demographically very similar, the teachers were well qualified, and the students had followed exactly the same mathematics approaches up to the age of thirteen, when my research began. At that time, the students at the two schools scored at the same levels on national mathematics tests. Then their mathematical pathways diverged.

The classrooms at Phoenix Park did look chaotic. The project-based approach meant a lot less order and control than in traditional approaches. Instead of teaching procedures that students would practice, the teachers gave the students projects to work on that *needed* mathematical methods. From the beginning of year 8 (when students started at the school) to three-quarters of the way through year 10, the students worked on open-ended projects in every lesson. The students didn't learn separate areas of mathematics, such as algebra or geometry, as English schools do not separate mathematics in that way. Instead, they learned "maths," the whole subject, every year. The students were taught in mixed-ability groups, and projects usually lasted for about three weeks.

At the start of the different projects, the teachers would introduce students to a problem or a theme that the students explored, using their own ideas and the mathematical methods that they were learning. The problems were usually very open so that students could take the work in directions that interested them. For example, in "Volume 216," the students were simply told that the volume of an object was 216 and they were asked to go away and think about what the object could be, what dimensions it could have, and what it could look like. Sometimes, before the students started a new project, teachers taught them mathematical content that could be useful to them. More typically though, the teachers would introduce methods to individuals or small groups when they encountered a need for them within the particular project on which they were working. Simon and Philip of year 10 described the school's maths approach to me in this way:

> S: We're usually set a task first and we're taught the skills needed to do the task, and then we get on with the task and we ask the teacher for help.
> P: Or you're just set the task and then you go about it. . . . You explore the different things, and they help you in doing that . . . so different skills are sort of tailored to different tasks.
> JB: And do you all do the same thing?
> P: You're all given the same task, but how you go

about it, how you do it, and what level you do it at changes, doesn't it?

The students were given an unusual degree of freedom in maths lessons. They were usually given choices between different projects to work on and they were encouraged to decide the nature and direction of their work. Sometimes the different projects varied in difficulty and the teachers guided students toward projects that they thought were suited to their strengths. During one of my visits to the classrooms, students were working on a project called "Thirty-six Fences." The teacher started the project by asking all of the students to gather round the board at the front of the room. There was a lot of shuffling of chairs as students made their way to the front, sitting in an arc around the board. Jim, the teacher, explained that a farmer had thirty-six individual fences, each of them one meter long, and that he wanted to put them together to enclose the biggest possible area. Jim then asked students what shapes they thought the fences could be arranged into. Students suggested a rectangle, triangle, or square. Jim asked, "How about a pentagon?" The students thought and talked about this. Jim asked them whether they wanted to make irregular shapes allowable.

After some discussion, Jim asked the students to go back to their desks and think about the biggest possible area that the fences could make. Students at Phoenix Park were allowed to choose whom they worked with, and some of them left the discussion to work alone, while most worked in pairs or groups of their choosing. Some students began by investigating different sizes of rectangles and squares, some plotted graphs to investigate how areas changed with different side lengths. Susan was working alone, investigating hexagons, and she explained to me that she was working out the area of a regular hexagon by dividing it into six triangles and she had drawn one of the triangles separately. She said that she knew that the angle at the top of each triangle must be sixty degrees, so she could draw the triangles exactly to scale using a compass and find the area by measuring the height.

I left Susan working and moved to sit with a table of boys. Mickey had found that the biggest area for a rectangle with perimeter 36 is 9 × 9. This gave him the idea that shapes with equal sides may give bigger areas and he started to think about equilateral triangles. Mickey seemed very interested in his work and he was about to draw an equilateral triangle when he was distracted by Ahmed, who told him to forget triangles since he had found that the shape with the largest area made of 36 fences was a 36-sided shape. Ahmed told Mickey to find the area of a 36-sided shape too, and he leaned across the table excitedly, explaining how to do this. He explained that you divide the 36-sided shape into triangles and all of the triangles must have a 1-cm base. Mickey joined in, saying: "Yes. Their angles must be 10 degrees!" Ahmed said: "Yes, but you have to find the height, and to do that you need the tan button on your calculator, T-A-N. I'll show you how. Mr. Collins has just shown me."

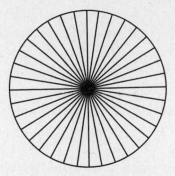

Mickey and Ahmed moved closer together, using the tangent ratio to calculate the area.

As the class worked on their investigations of thirty-six fences, many of the students divided shapes into triangles. This gave the teacher the opportunity to introduce students to trigonometric ratios. The students were excited to learn about trig ratios as they enabled them to go further in their investigations.

At Phoenix Park, the teachers taught mathematical methods to help students solve problems. Students learned about statistics and probability, for example, as they worked on a set of activities called "Interpreting the World." During that project they interpreted data on college attendance, pregnancies, football results, and other issues of interest to them. Students learned about algebra as they investigated different patterns and represented them symbolically; they learned about trigonometry in the "Thirty-six Fences" projects and by investigating the shadows of objects. The different projects were carefully chosen by the teachers to interest the students and to provide opportunities for learning important mathematical concepts and methods. Some projects were applied, requiring that students engage with real-world situations; other activities started with a context, such as thirty-six fences, but led into abstract investigations. As students worked, they learned new methods, they chose methods they knew, and they adapted and applied both. Not surprisingly, the Phoenix Park students came to view mathematical methods as flexible problem-solving tools. When I interviewed Lindsey in the second year of the school, she described the maths approach: "Well, if you find a rule or a method, you try and adapt it to other things. When we found this rule that worked with the circles, we started to work out the percentages and then adapted it, so we just took it further and took different steps and tried to adapt it to new situations."

Students were given lots of choices as they worked. They were allowed to choose whether they worked in groups, in pairs, or alone. They were often given choices about activities to work on and they were always encouraged to take problems in directions that were of interest to them and to work at appropriate levels. Most of the students liked this mathematical freedom. Simon told me: "You're able to explore. There's not many limits and that's more interesting." Discipline was very relaxed at Phoenix Park, and students were also given a lot of freedom to work or not work.

Amber Hill School

At Amber Hill School the teachers used the traditional approach that is commonplace in England and in the United States. The teachers began lessons by lecturing from the board, introducing students to mathematical methods. Students would then work through exercises in their books. When the students at Amber Hill learned trigonometry, they were not introduced to it as a way of solving problems. Instead, they were told to remember

and they practiced by working through lots of short questions. The exercises at Amber Hill were typically made up of short contextualized mathematics questions, such as:

> **Helen rides a bike for 1 hour at 30 km/hour and 2 hours at 15 km/hour. What is Helen's average speed for the journey?**

Classrooms were peaceful and quiet at Amber Hill and students worked quietly, on task, for almost all of their lessons. Students always sat in pairs and they were generally allowed to converse quietly—usually checking answers with each other— but not encouraged to have mathematical discussions. During the three years that I followed the students as they progressed through school, I learned that the students worked hard but that most of them disliked mathematics. The students at Amber Hill came to believe that maths was a subject that only involved memorizing rules and procedures. As Stephen described to me: "In maths, there's a certain formula to get to, say from *a* to *b,* and there's no other way to get to it. Or maybe there is, but you've got to remember the formula, you've got to remember

it." More worryingly, the students at Amber Hill became so convinced of the need to memorize the methods they were shown
that many of them did not see any place for thought. Louise, a
student in the highest group, told me: "In maths you have to
remember. In other subjects you can think about it."

Amber Hill's approach stood in stark contrast to Phoenix
Park's. The Amber Hill students spent more time on tasks, but
they thought maths was a set of rules that needed to be memorized, and few of them developed the levels of interest the
Phoenix Park students showed. In lessons the Amber Hill students were often successful, getting lots of questions right in
their exercises, not by understanding the mathematical ideas
but by following cues. For example, the biggest cue telling students how to answer a question was the method they had just
had explained on the board. The students knew that if they
used the method they had just been shown, they were probably
going to get the questions right. They also knew that when they
moved from exercise A to exercise B, they should do something
slightly more complicated. Other cues included using all the lines
given to them in a diagram and all the numbers in a question; if
they didn't use them all, they thought they were doing something
wrong. Unfortunately, the same cues were not present in the
exams, as Gary told me, when describing why he found the
exams hard: "It's different, and like the way it's there, like, not
the same. It doesn't, like, tell you it—the story, the question; it's
not the same as in the books, the way the teacher works it out."
Gary seemed to be suggesting, as I had seen in my observations, that the story or the question in their books often gave
away what they had to do, but the exam questions didn't. Trevor
also talked about cues when he explained why his exam grade
hadn't been good: "You can get a trigger, when she says like
'simultaneous equations' or 'graphs,' or 'graphically.' When they
say like—and you know, it pushes that trigger, tells you what to

do." I asked him, "What happens in the exam when you haven't got that?" He gave a clear answer: "You panic."

In England all students take the same national examination in mathematics at age sixteen. The examination is a three-hour, traditional test made up of short mathematics questions. Despite the difference in each school's approach, the students' last-minute preparation for the examination was fairly similar as both schools gave students past examination papers to work through and practice. At Phoenix Park the teachers stopped the project work a few weeks before the examination and focused upon teaching any standard methods that students may not have met. They spent more time lecturing from the board, and classrooms looked similar (briefly) to those at Amber Hill.

Many people expected the Amber Hill students to do well on the examinations, as their approach was meant to be examination oriented, but it was the Phoenix Park students who achieved significantly higher examination grades. The Phoenix Park students also achieved higher grades than the national average, despite having started their school at significantly lower levels than the national average. The examination success of the students at Phoenix Park surprised people in England and the research study was reported in all of the national newspapers. People believed that a project-based approach would result in great problem solvers, but they had not thought that an approach that was relaxed and project based with no "drill and practice" could also result in higher examination grades.

The Independent

Trendy teachers are top in maths

WENDY BERLINER   mixed ability groups in every maths lesson. There was ver lit-

The Times

Progressive methods 'best for maths'

Progressive maths teaching led to better results   **by Diane Hofkins**   approach, did not apply themselves and would have preferred textbooks.

The Guardian

They've got the right formula

These were not quite the headlines I would have chosen, but the approach was gaining rightful attention.

The Amber Hill students faced many difficulties in the examination, which they were not expecting, as they had worked so hard in lessons. In class the Amber Hill students had always been shown methods and then practiced them. In the examination they needed to choose methods to use and many of them found that difficult. As Alan explained to me: "It's stupid really, 'cause when you're in the lesson, when you're doing work— even when it's hard—you get the odd one or two wrong, but most of them you get right and you think, 'Well, when I go into the exam, I'm gonna get most of them right,' 'cause you get all your chapters right. But you don't." Even in the examination questions when it was obvious which methods to use, the Amber Hill students would frequently confuse the steps they had learned. For example, when the Amber Hill students answered a question on simultaneous equations, they attempted to use the standard procedure they had been taught, but only 26 percent of the students used the procedure correctly. The rest of the students used a confused and jumbled version of the procedure and received no credit for the question.

The Phoenix Park students had not met all of the methods they needed in the examination, but they had been taught to solve problems and they approached the examination questions in the same flexible way as they approached their projects— choosing, adapting, and applying the methods they had learned. I asked Angus whether he thought there were things in the exam that they hadn't seen before. He thought for a while and said: "Well, sometimes I suppose they put it in a way which throws you. But if there's stuff I actually haven't done before, I'll try and make as much sense of it as I can, try to understand it, and answer it as best as I can, and if it's wrong, it's wrong."

The Phoenix Park students didn't only do better on the examinations. As part of my research, I investigated the usefulness of the approach to students' lives. One way I measured this was by giving a range of assessments to students over the three years that were designed to assess students' use of mathematics in real-world situations. In the "architectural activity," for example, students had to measure a model house, use a scale plan, estimate, and decide upon appropriate house dimensions. The Phoenix Park students outperformed the Amber Hill students on all of the different assessments. By the time I was completing the research, most of the students had jobs in the evenings and on weekends. When I interviewed the students at both schools about their use of mathematics outside school, there were stark differences. All forty of the Amber Hill students that I interviewed said that they would never, ever make use of their school-learned methods in any situation outside school. As Richard told me: "Well, when I'm out of school, the maths from here is nothing to do with it, to tell you the truth. . . . Most of the things we've learned in school we would never use anywhere." The Amber Hill students thought school mathematics was a strange sort of code that you would use in one place—the mathematics classroom—and they developed the idea that their school mathematics knowledge had boundaries or barriers surrounding it, which kept it firmly within the mathematics classroom.[2,3]

At Phoenix Park, the students were confident that they would utilize the methods they learned in school. They gave me examples of their use of school-learned mathematics in their jobs and lives. Indeed, many students' descriptions suggested that they had learned mathematics in a way that transcended the boundaries that generally exist between the classroom and real situations.[4]

Mathematics for Life

Some years later I caught up with the ex-students from Amber Hill and Phoenix Park. By then they were about twenty-four years old. We talked about the usefulness of the math teaching they had experienced. I had frequently been asked about the future of the students after they had left their schools and so I decided to find out. I sent surveys to the ex-students' addresses and followed up the surveys with interviews. As part of the survey I asked the young people what jobs they were doing. I then classified all the jobs and put them onto a scale of social class, which gives some indication of the professionalism of their jobs and their salaries. This showed something very interesting. When the students were in school, their social class levels (determined from parents' jobs) had been equal. Eight years after my study, the Phoenix Park young adults were working in more highly skilled or professional jobs than the Amber Hill adults, even though the school achievement range of those who had replied to the surveys from the two schools had been equal. Comparing the jobs of the children to their parents, 65 percent of the Phoenix Park adults had moved into jobs that were more professional than their parents', compared with 23 percent of Amber Hill adults. Fifty-two percent of the Amber Hill adults were in less professional jobs than their parents, compared with only 15 percent of the Phoenix Park adults. At Phoenix Park, there was a distinct upward trend in careers and economic well-being. At Amber Hill, there was not, which is especially noteworthy considering that Phoenix Park was in a less prosperous area.

When I traveled back to England to conduct the follow-up interviews, I contacted a representative group from each school, choosing young adults with comparable examination grades. In interviews the Phoenix Park adults communicated a positive approach to work and life, describing the ways they used the problem-solving approach they had been taught in their mathematics classrooms to solve problems and make sense of mathematical situations in their lives. Adrian, who had studied economics at a university, told me that "you often get lots of stuff where there will be graphs of economic situations in countries and stuff like that. And I would always look at those very critically. And I think the maths that I've learned is very useful for being able to actually see exactly how it's being presented, or whether it's being biased."

When I asked Paul, who was a senior regional hotel manager, whether he found the mathematics he had learned in school useful, he said that he did: "I suppose there was a lot of things I can relate back to maths in school. You know, it's about having a sort of concept, isn't it, of space and numbers and how you can relate that back. And then, okay, if you've got an idea about something and how you would then use maths to work that out. . . . I suppose maths is about problem solving for me. It's about numbers; it's about problem solving; it's about being logical."

Whereas the Phoenix Park young adults talked of maths as

a problem-solving tool, and they were generally very positive about their school's approach, the Amber Hill students could not understand why their school's mathematics approach had prepared them so badly for the demands of the workplace. Bridget spoke sadly: "It was never related to real life, I don't feel. I don't feel it was. And I think it would have been a lot better if I could have seen what I could use this stuff for . . . because then it helps you to know *why*. You learn *why* that is that, and *why* it ends up at that. And I think definitely relating it to real life is important."

Marcos was also puzzled as to why the school's maths approach had seemed so removed from the students' lives and work: "It was something where you had to just remember in which order you did things, and that's it. It had no significance to me past that point at all—which is a shame. Because when you have parents like mine who keep on about maths and how important it is, and having that experience where it just seems to be not important to anything at all really. It was very abstract. And it was kind of almost purely theoretical. As with most things that are purely theoretical, without having some kind of association with anything tangible, you kind of forget it all."

My book of the study of the approaches at Phoenix Park and Amber Hill, *Experiencing School Mathematics,* won a national book award in England and has been read by thousands of British, American, and other readers. Many teachers have contacted me saying that they would like to teach through a problem-solving approach similar to Phoenix Park's but they cannot because of lack of support from departments, administration, and parents. But now is the time to make changes, to grasp the optimism and enthusiasm existing around math change, and to make decisions that are informed by research about the best math education we can provide to children.

The two teaching approaches I have reviewed were the subject of comprehensive research studies and, although they were conducted in different countries, the findings pointed to the same conclusion: students need to be actively involved in their learning and they need to be engaged in a broad form of mathematics—using and applying methods, and representing and communicating ideas. As a professor at Stanford, I am frequently contacted by teachers, district officials, and parents wanting to know which curricula are good to use in math classrooms. This is a question I find difficult to answer, as I strongly believe that teachers are the most important part of an approach and it is hard to recommend a book or a curriculum approach without knowing how a teacher is using it. But it is the case that some books have been written to involve students more actively and some books include very good mathematics problems. In appendix B, I list some of these books, including those that I personally regard to be of a high quality.

4 / Taming the Monster

New Forms of Testing That
Encourage Learning

D oes it matter that American children are tested more than they ever used to be? Does it matter that they are tested more than students in the rest of the world? And does it matter that the tests used in America are rejected by most other countries? The answer to all of these questions is, of course, yes. America is out of sync with the rest of the world, not because the United States has a better system of testing, as one might expect from such a well-resourced country, but because the testing system in the United States is disastrous. Students are overtested to a ridiculous degree, and the tests that are used are damaging—to schools and teachers and, most important, to the health, hearts, and minds of students.

Alfie Kohn, author of numerous books on education, writes that "Standardized testing has swelled and mutated, like a creature in one of those old horror movies, to the point that it now threatens to swallow our schools whole."[1] The testing move-ment has indeed swallowed many of our schools whole, and it is time to do something about it. Fortunately, there is a new form of assessment that is so powerful and so effective, not only at finding out what students know but also at diagnosing and improving learning, that it has energized a whole new movement. It is called "assessment for learning" and it has been shown to have an effect on learning that, if implemented, would raise America's ranking in international comparisons from the middle of the pack to a place in the top five. In England, a booklet called *English Inside the Black Box*[2] (which detailed the new assessment methods) sold tens of thousands of copies in a very short space of time. The schools that used the methods and were part of a careful research study[3] significantly improved their students' achievement and attitudes toward work.

Assessment for learning aims to create self-regulating learners—learners who have the knowledge and power to monitor their own learning. The testing monster has the biggest impact on math classrooms and math learning, and we know that math has the power to crush the spirit of young people in ways that no other subject can. How wonderful, then, that there is a new way of assessing math (and other subjects) that doesn't terrorize chil-dren, doesn't distort the curriculum, and propels students to higher levels of learning. If it sounds too good to be true, please read on.

What Is Wrong with What We Have Now?

Students in American schools are subject to narrow standard-ized tests in mathematics—as well as other subjects—from a very young age. Although most countries in the world test students by posing questions that students respond to in writing and that are graded by trained experts, America uses multiple-choice tests that are graded by machines. It is hard to find a single multiple-choice question used in Europe—in any national assessment, in any subject, at any level, in any country—yet almost all of America's test questions are of a multiple-choice format. I will speculate that the reasons other countries do not use multiple-choice questions in their examinations are fourfold. First, they want to assess understanding, which includes the thinking that children do and express in words, numbers, and symbols. What children choose to put on paper is the best indicator of what they understand, not their choice of one out of four options, none of which may mean anything to them. Second, multiple-choice testing is known to be biased—particularly for ethnic minority students. There has been mixed evidence on the bias of multiple-choice tests against girls, but we do know that girls do less well on tests such as the SATs, designed to predict college performance, but then outperform boys to a significant degree in college.[4] Third, undergoing timed multiple-choice tests causes anxiety and contributes to the stressed-out nation of schoolchildren that America now has.[5] Fourth, the best thing that multiple-choice tests show is a student's ability to complete multiple-choice tests. Some students are good at that type of test taking and do well on those tests, while other students, including those who are highly intelligent and knowledgeable, do badly. Martin Luther King, for example, one of the country's most influential writers, scored in the lowest 10 percent of students on both the math and verbal sections of the Graduate Record Examination, despite being such a brilliant student that he entered college at the age of sixteen.[6] Knowing that a student has done well, or badly, on a multiple-choice test does not tell you much, if anything, about how well they will handle more advanced material or solve complex problems in the workplace.

In addition to a reliance on multiple-choice formats, the mathematics tests used in most states across America are extremely narrow. They do not assess thinking, reasoning, or problem solving, all of which are at the core of mathematics; instead, they assess the simple use of procedures, completed under timed conditions. Procedures are important, of course, but only if they can be used to solve problems. What is the point of knowing procedures if students don't know when they should use them, or how to apply them to complex problems? One of the most important principles of good testing is that it assesses what is important. The tests that predominate in America do not. The worst of this is not that the tests provide little information but that they have a huge and damaging impact on what is taught in schools. Almost every mathematics teacher in America will tell you that the pressure to prepare students for standardized tests harms their teaching and their students' learning. In mathematics the teachers have to focus upon knowledge that can be tested rather than knowledge that is important for work or for life. Students also suffer from an extreme narrowing of the curriculum. In North Carolina, for example, the pressure to perform on high-stakes tests resulted in widespread decline in the teaching of science, social studies, physical education, and the arts.[7]

Many people in America think that achievement is low because teachers are poorly qualified or lack knowledge. At Stanford I teach future math teachers who are highly qualified and committed. They are people with math degrees from some of the most prestigious universities in the world, people who have turned down much more lucrative careers so they can teach children. They know how important it is that students learn to think and reason, and that they are willing and able to solve complex problems. But at the same time the teachers often feel unable to spend any time teaching students to do these things while they are so constrained by California's standardized tests. Alfie Kohn quotes from a teacher who had to stop one of her most successful teaching activities because of the high-stakes tests. She used to ask her middle school students to become experts in a subject, developing their researching and writing skills. This was an experience that her students remembered for years, that they looked back on as a highlight of school, but that she was forced to end. As Kohn writes: "Within each classroom 'the most engaging questions kids bring up spontaneously—"teachable moments"—become annoyances.' Excitement about learning pulls in one direction; covering the material that will be on the test pulls in the other."[8] The elimination of powerful learning experiences because they cannot be reduced to testable knowledge is damaging education in America.

The tests used in America are particularly harmful for children of low income and for English language learners (ELL students). America is currently one of the most unequal countries in the world, and the gap between the educational achievement of the wealthy and the poor is vast. The No Child Left Behind Act, brought in by President Bush's administration, made standardized testing compulsory across the country, with the supposed aim of eradicating inequalities. The irony and the tragedy of this is that the testing movement has made inequalities worse.

There is also evidence that the standardized questions test language as much as they do mathematics. In California in 2004, there was a staggering correlation of 0.932 between students' scores on the mathematics and language arts sections of the tests used. Correlations this high between two tests tell us that the tests are assessing virtually the same thing. Even the same test taken twice at different times might not give us such a high correlation. We should expect some correlation between people's scores on mathematics and language arts tests, as good students sometimes do well in both, but a correlation of 0.932 is ridiculously high. And since the language arts questions have no mathematics in them but the mathematics questions use unnecessarily complex language, there can be only one reason that they are so highly correlated: the mathematics tests are really language tests. These are the tests that are being used to judge students' mathematical understanding and the tests that are driving the curricula in schools.

We are entering a new era of Common Core math and associated assessments. We will not be free from standardized testing, but the new assessments do promise to fundamentally change the way children are tested, replacing the dreaded multiple-choice items with free-response items in which students explain their thinking. The new assessments come from two different assessment companies, the Parthership for Assessment of Readiness for College and Careers (PARCC) and Smarter Balanced Assessment Consortium, which are promising to assess problem solving rather than procedure repetition. It is too early to say whether the assessment companies fulfill the promises made, but we have to hope that they do, because assessment drives instruction.

When I was researching Railside High School,[9] with its outstanding mathematics department, I met Simon from Nicaragua, who had arrived in this country as a young boy. Simon told me that elementary school was a time of constant failure, as he couldn't understand what the teachers were saying. Since then, though, he had attended wonderful schools and was excelling in all of his subjects. Always happy and smiling, Simon was one of those students whom teachers love to teach. He told us that the teachers at Railside convinced him that he was smart, and he started to believe in himself and achieve. When I met Simon, he told me that he loved math. In different assessments, including ours from Stanford, he performed extremely well. Despite all of this, he had not done well on the standardized test administered by the state of California— at that time the SAT-9 tests. The reason for this had little to do with mathematics understanding but instead with the unfamiliar language and contexts used in the test and the strange format of the questions.[10] Many of the ELL students at Railside underperformed on the tests and the Railside teachers were put under pressure to spend less time on important mathematics and more time on training students to complete a multiple-choice test. It was very sad to see the teachers and students spending their time in these ways, especially as they were being pulled away from important mathematical activities and learning opportunities.

In addition to the harm done by inadequate tests that do not assess mathematical understanding, additional punishment is dealt to many of the students by the harsh and comparative reporting of scores. When students are sent a label telling them where they stand compared to other students, rather than where they stand in their learning of mathematics, it offers no helpful information and is harmful to many students. A test should communicate to students *what* they have learned and how much they have learned *over a period of time.* At the very least, tests should set out for students what they know and do not know so that they give students information to work with.[11, 12] Knowing that you are doing worse than others is simply demoralizing, as Simon discovered. Simon had learned a huge amount at Railside and worked extremely hard, but the odds were stacked against him when taking a test using convoluted language and strange contexts. He did not do well on the test, but instead of finding out which questions he did badly on, so that he could work on those, or how much he had improved over the past year (a great deal), he received a label telling him where he was ranked compared to others in the country. This practice was especially unfair as some states allow calculators in the tests but his state (California) did not. Simon was simply and devastatingly told that he was "below average." The label that Simon's parents received in the mail caused him to question his ability despite all of his achievements at school. As he told us: "My parents, they saw in the SAT-9 graph thing I was *below average* in the majority of the things, and especially math. I was like *below average*. Right there. The thing is like *below average*, you want it to be a little bit above average."

I asked him whether that affected how he thought of his

abilities as a mathematics learner. He said that it did: "You tried so hard and then suddenly they give you a paper where it says you're *below average* and you're like, 'What? I did so much work.'"

Simon had reasonably assumed that the result he was given should tell him something about how hard he had worked or what he had learned in mathematics, but it did neither of these things. Testing and reporting measures such as those experienced by Simon can *create* low-achieving students, crushing students' confidence and giving them an identity as a low achiever.

Claude Steele, an eminent psychologist, has demonstrated the damage caused by "stereotype threat."[13] He found that when students took a math test in which they recorded their gender at the start of the test, women underperformed. In his earlier studies, he found that women underperformed if they were reminded of gender differences in math at the start of the test, but in subsequent experiments he found that people did not need to be reminded of gender differences because stereotypes are always "in the air," and a simple recording of gender on a test produced gender differences for women in math.[14] Researchers have shown the same phenomenon for any group that feels that they could be lower performers. This includes white men given a golfing test alongside African American men, who the white men believe will be superior golfers.[15] Educational research—a field that often produces conflicting results—shows remarkable consistency on this issue. If you tell students they are low achievers, they achieve at a lower level than if you do not. At the time that Simon received his label, almost half of all students in the state were told that they were below average. At the time, students were told that they were "basic," "below basic," or even "far below basic." California is now proposing to use levels instead of words. What impact, I wonder, do those who design these systems think that they will have upon students' confidence and their future mathematics achievement? Research tells us that confidence in one's ability to succeed in mathematics is an intrinsic part of success and motivation. The labels the students in America receive tell many of them that there is no point in trying.

The ways in which assessment reporting can affect students' confidence as learners was illustrated poignantly by a ten-year-old student in England reporting on the standardized tests (called SATs) that she was about to take. She said: "I'm really scared about the SATs. Mrs O'Brien came and talked to us about our spelling and I'm no good at spelling and David [the class teacher] is giving us times tables tests every morning and I'm hopeless at times tables, so I'm afraid I'll do the SATs and I'll be a nothing." When the interviewer tried to convince her that she could never be a "nothing," no matter what happened on the tests, the young girl insisted that the tests would make her one. The researchers reported that even though she was "an accomplished writer, a gifted dancer and artist, and good at problem solving," the tests made her feel as though she was an "academic non-person."

As if the requirements that students take low-quality tests and suffer crude and demoralizing labeling were not bad enough, many teachers feel the need to coach for these tests by using similar tests in their classrooms. Many of the math teachers I have met in the United States give students a test every week, and at the end of each chapter in their textbook, that are often replicas of chapter questions. Some teachers test students more frequently, and students end up spending almost as much time taking tests as they do learning new material. Teachers rarely do anything with the tests other than score them, return the scores, then move on to the next section of work. Dylan Wiliam, an international expert on assessment, has likened this practice to a pilot of a plane flying off into the distance and hoping to land in New York. After a certain amount of time the pilot lands in the nearest airport and asks, "Did I arrive in New York?" Whether they did or not, all the passengers would be instructed to get off so that the plane could then fly off to its next destination. Teachers in America typically do the equivalent—they teach a chapter or unit of work and at the end they give a test. They then move on to another chapter, whether students are still with them or not. Giving a test at the end of the work, with no intermediate indicators of understanding, means that teachers can do little with the information they gain from tests, apart from going over the answers to badly answered questions. Students who are confused the first time are usually confused again. With the pressure of an overfilled curriculum and a lack of knowledge about more effective assessment methods, teachers do not spend time using assessments *to improve learning*.

Teachers often treat the reporting of test scores just as badly, typically giving students only enough information to compare their performance with others. Such feedback is minimally helpful to some and has a negative impact on many. When students are given only a percentage or grade, they can do little else besides compare it to others around them, with half or more deciding that they are not as good as others. This is known as "ego feedback," a form of feedback that has been found to be damaging to learning. A large review of research from hundreds of studies showed that such feedback has a negative impact on performance in 38 percent of cases. If students are given scores that tell them they are below others in the class, it can damage self-esteem to the extent that students give up on math or take on the identity of an underperforming student. It is really not helpful to be told that you have scored 65 percent, unless you know how to do better on the remaining 35 percent of the questions you answered incorrectly. Deevers[16] found that students who were not given scores but instead given positive constructive feedback were more successful in their future work.

Unfortunately, he also found that teachers gave less and less constructive feedback as students got older. When he explored the relationship between teachers' assessment practices and student attitudes and beliefs about mathematics, he found that students' belief in the power to improve their own learning, and their motivation to learn, declined steadily from fifth to twelfth grade.

The standardized tests such as those imposed upon the country by the No Child Left Behind Act, as well as those typically used inside math classrooms in the United States, do not give any information that helps students improve their learning. Instead, they compare students to one another. There should be some large-scale tests in use in schools—tests that are designed to give a measure of a students' performance at the end of a course, or to give national or state comparisons—but these do not need to be as low-quality as those administered by most states and they should not be multiple-choice tests. The AP examinations administered at the end of AP courses are a good example of a better-quality assessment that can be administered large scale and that do not harm the curriculum or learning. Inside classrooms there is room for even more improvement, and the "assessment for learning" approach has been designed to revolutionize the way assessment is used inside classes.

Assessment for learning is a form of assessment that gives useful information to teachers, parents, and others, but it also empowers students to take charge of their own learning. Assessment for learning tells students where they are in their learning path, where they could be, and what they need to do to get there. It gives learners the motivation *and the power* to regulate and improve their own learning, and there are particular ways in which parents can be involved with this new approach to improve their children's life chances.

Assessment for Learning

Assessment for learning is based upon the principle that students should have a full and clear sense of what they are learning, of where they are in the path toward mastery, and of what they have to do to become successful. Students are given the knowledge and tools to become self-regulatory learners, so that they are not dependent on following somebody else's plans, with little awareness of where they are going or what they might be doing wrong. It may seem obvious that learners should be clear about what they are learning and what they need to do to be successful, but in most mathematics classrooms the students have very little idea. I have visited hundreds of classrooms and stopped at students' desks to ask them what they are working on. In traditional classrooms, students usually tell me what page they are on, or what exercise they are working through. If I ask them, "but what are you actually doing?" they say things like "Oh, I'm doing number three." Students are usually able to tell you the titles of chapters they are working on, but they really do not have a clear sense of the mathematical goals they are pursuing, the ways that the exercises they work through are linked to the bigger goals they are pursuing, or the differences between more and less important ideas. This makes it very difficult for students, or parents, to do anything to improve children's learning. Mary Alice White, who was a professor of psychology at Columbia University, had likened the situation to workers on a ship who may be given small tasks to do, and complete them each day without having any idea of where the ship is heading or the voyage they are undertaking.

The first part of the assessment-for-learning approach involves communication about what is being learned and where students are going. The second part involves making individual students aware of where they are in the path to success, and the third part involves giving them clear advice about how to become more successful. The approach is called "assessment *for* learning" rather than "assessment *of* learning" because it is designed to promote learning, and all the information that is gained from assessment is made helpful to individual learners to propel them to greater levels of success.

So how is this achieved, and how does assessment for learning look different from traditional assessment in the classroom? First of all, students are made aware of what they are, should be, and could be learning through a process of self- and peer assessment. Teachers set out mathematical goals for students, not a list of chapter titles or tables of contents, but details of the important ideas and the ways they are linked. For example, students might be given a range of statements that describe the understanding they should have developed during a piece of work, with statements such as "I have understood the difference between mean and median and know when each should be used." The statements are clear for students to understand, and they communicate to them what they should be understanding from a piece of work. Students then assess their own or their peers' work against the statements. By assessing work against statements and deciding what they have understood, students come to understand the goals of lessons much more clearly. In reviewing the goals of a lesson, week, or unit of work, students start to become aware of what they should be learning and what the big ideas are. Parents can also review the criteria and become more knowledgeable about the ideas and knowledge students are working toward.

In studies of self-assessment in action, researchers have found that students are incredibly perceptive about their own learning, and they do not over- or underestimate it. They carefully consider goals and decide where they are and what they do and do not understand. In peer assessment, students are asked to judge each other's work, again evaluating the work against clear criteria. This has been shown to be very effective, in part because students are often better able to hear criticism from their peers than from a teacher, and peers usually communicate in easily understood ways. It is also an excellent opportunity for students to become aware of the criteria against which they too are being judged. One way of managing peer assessment is to ask students to identify "two stars and a wish"—they select two things done well and one area to improve in their peers' work. When students are frequently asked to consider the goals of their learning, for themselves or their peers, they become very knowledgeable about what they are meant to be learning, and this makes a huge difference.

Evidence of the power of making students aware of what they are meant to be learning came from a very careful study conducted by two psychologists: Barbara White and John Frederiksen.[17] They worked with twelve classes of thirty students learning physics. Each class was taught a unit on force and motion, with students in each class divided into experimental and control groups. The control group used some periods of time each lesson to discuss the work, while the experimental group spent the same time engaging in self- and peer assessment, considering assessment criteria. The results were dramatic, with the experimental group outperforming the control group on three different assessments. The greatest gains were made by the students who had previously been the lowest-achieving students. After they spent time considering criteria and assessing themselves and their peers against them, the previously low achievers began to achieve at the same levels as the highest achievers. Indeed, the seventh grade students who reflected on the criteria scored at higher levels than AP physics students on tests of high school physics. White and Frederiksen concluded that the students were previously unsuccessful not because they lacked ability but because they had not really known what they were meant to be focusing on.

When students are required to be aware of what they are

learning and to consider whether they understand it, this also gives important information to the teacher. For example, consider an assessment-for-learning activity called "traffic lighting." In some versions of this, students are asked to put a red, orange, or green sticker on their work to say whether they understand new work well, a little, or not at all. In other versions, the teachers give students three paper cups: one green, one yellow, and one red. If a student feels the lesson is going too fast, he or she shows the yellow cup; those who need the teacher to stop can show the red cup. At first researchers found that students were reluctant to show a red cup, but when teachers asked someone who was showing a green cup to offer an explanation, students became more willing to show a red cup when they were not understanding something. Other teachers used the traffic lighting to group students, sometimes having the greens and yellows work together to deal with problems between themselves, while the red pupils were helped by the teacher to deal with deeper problems. Most important, the students themselves were being asked to think about what they knew and could do and what they needed more help on. This helps students and teachers tremendously, as teachers get feedback on their teaching in real time, rather than at the end of a unit or piece of work when it is too late to do anything about it, and they are able to deliver the most helpful information to students at an appropriate pace.

Methods of self- and peer assessment serve the purpose of both teaching students about the goals of the work and the nature of high-quality work and giving them information on their own understanding. For teachers, these methods also give critical information on student understanding that can help them assist individuals in the best possible way and improve their own teaching. But, as Paul Black and Dylan Wiliam have noted, these new methods require fundamental changes in the behavior of both students and teachers. Students need to move from being passive learners to being active learners, taking responsibility for their own progress, and teachers need to be willing to lose some of the control over what is happening, which some teachers have described as scary but ultimately liberating. As Robert, a teacher at Two Bishops School in England, said when reflecting on his new assessment-for-learning approach: "[What it] has done for me is made me focus less on myself but more on the children. I have had the confidence to empower the students to take it forward."[18]

The third part of the assessment-for-learning approach—after making students aware of what needs to be learned and how they are doing—involves helping students know *how* to improve, which is best achieved through diagnostic feedback. Psychology professors Maria Elawar and Lyn Corno[19] trained eighteen sixth-grade teachers in three schools in Venezuela to give constructive written feedback in response to mathematics homework, instead of the scores they normally gave. The teachers learned to comment on errors (giving specific suggestions about how to improve) and to give at least one positive remark about each student's work. In an experimental study, half the students received homework grades as usual (just scores), and half received constructive feedback. The students receiving the constructive feedback learned twice as fast as the control-group students, the achievement gap between male and female students was reduced, and students' attitudes toward mathematics became significantly more positive.

In another interesting study, Ruth Butler, a professor of education at the Hebrew University of Jerusalem,[20] compared three ways of giving feedback to different groups of students. One group received grades; one group received comments on their work, saying whether carefully explained criteria were matched; and one group received both grades and comments. What the researchers found was that the group receiving comments increased their performance significantly, whereas those who received grades did not. More surprising, perhaps, those who received both grades and comments did as poorly as those who only received grades. It turned out that those who received both grades and comments only focused upon the grades they received, which served to override any comments. Diagnostic, comment-based feedback is now known to promote learning, and it should be the standard way in which students' progress is reported. Grades may be useful for communicating where students are in relation to one another, and it is fine to give them at the end of a semester or term, but if they are given more frequently than that, they will reduce the achievement of many. More typical reporting should be made up of feedback on the mathematics that students have learned, clear insights into the mathematics students are working toward, and advice on how to improve.

One assessment expert puts it simply: "Feedback to learners should focus on what they need to do to improve, rather than on how well they have done, and should avoid comparison with others."[21] This makes perfect sense. Coaches tell athletes how to be better; they don't just tell them a grade. So why should teachers just say, in so many words, "You are a low achiever"? As Royce Sadler, a professor of higher education, says, "The indispensable conditions for improvement are that the student comes to hold a concept of quality roughly similar to that held by the teacher, is continuously able to monitor the quality of what is being produced during the act of production itself, and has a repertoire of alternative moves or strategies from which to draw at any given point."[22] The beauty of the assessment-for-learning approach is that it not only gives critical information to learners but that it also gives the same important information to teachers, helping them gear their teaching to their students' needs.

The main aim of assessment for learning is the improvement of *classroom* assessment, but the methods have also been used to improve wide-scale assessments such as those used at national levels. This is important because we know that teachers tend to mimic state and national assessments in their own classrooms. In Queensland, Australia, for example, the government recognized the need for good assessments that could inform and improve learning rather than traditional assessments that simply ranked students, and they designed a system to ensure objectivity. The system was instigated in 1971 and has continued to evolve and be improved upon since then. It involves students taking two assessments: a school-based "course of work" (in which teachers assess their students' work in class, which is moderated by a committee) and a test of core skills. The test is used for the purpose of comparing performance across subjects. If there is a discrepancy between performance on the school work and the skills test, then the former takes precedence. Importantly, the school-based work offers opportunity for good assessment tasks, criteria that students can consider as they work, and diagnostic teacher feedback, combined with good, reliable information for parents and others. Other countries are now developing large-scale assessments that encourage learning but that can also be used to give unbiased, objective measurements of students' learning through well-developed systems to ensure that graders are following the same standards.

Good assessments are critical at national and state levels as well as in classrooms. They should help students know what they are learning, provide students with the opportunity to show their understanding thoroughly, show students (and others) where they are in their learning, and give feedback on how to improve. Importantly, students should be given information that refers to the content area being assessed, not to other students. Assessment for learning transforms students from passive receivers of knowledge to active learners who regulate their own progress and knowledge and propel themselves to higher levels of understanding. It also broadens teachers' testing strategies, encouraging them to focus less on simple tests and more on the broad ways in which they can monitor student learning, including class work and student discussions and presentations. National and state assessments should lead the way in educating teachers about good assessments, and these powerful new methods should be in place in all classrooms in America.

Dostoensky

5 / Stuck in the Slow Lane

How American Grouping Systems
Perpetuate Low Achievement

Mixed grouping → higher achievement

Whether or not to group students by prior attainment (often called ability) is one of the most controversial topics in education. Many parents support ability grouping because they want their high-achieving, motivated children to be working with similar children. This is completely understandable and makes perfect sense. But we know from several international studies that countries that reject ability grouping—nations as varied as Japan and Finland—are among the most successful in the world, whereas countries that employ ability grouping, such as America, are among the least successful. Is ability grouping part of the problem of America's low achievement? And if it is, how can it be the case, when the idea seems to make so much sense?

In the Third International Mathematics and Science Study (TIMSS) researchers collected a wide range of data on eighth-grade students in thirty-eight countries.[1] The achievement results,

US most tracking

with the United States coming in nineteenth place, received a huge amount of attention and prompted widespread concern. But analysts of the TIMSS database also uncovered some interesting facts about ability grouping. For example, in a study of achievement variability, the United States was found to have the greatest amount of variability between classes—that is, the United States had the most tracking.[2] The highest-achieving country in the group, Korea, was the country with the least tracking and the most equal grouping. The United States also had the strongest links between achievement and socioeconomic status (SES), a fact that has been attributed to the widespread use of tracking in American schools. The high achievement of countries that do not use ability grouping and the low achievement of those that do was also a finding of the Second International Mathematics and Science Study (SIMSS). This caused leading analysts to conclude that countries that leave grouping to the *latest* possible moment or that use the *least* amount of grouping by ability are those with the highest achievement.

Some have argued that the high degree of tracking in America is a reflection of our desire to sort students at an early stage in order to find and focus upon the high-achieving students. But such an approach has some serious flaws, including the difficulty of identifying students correctly when children develop at different rates, and the creation of highly unequal schooling, which is one of the results identified by the international studies. In Japan, by contrast, the main priority is to promote high achievement for all, and teachers refrain from prejudging achievement, instead providing all students with complex problems that they can take to high levels. Japanese educators are bemused by the Western goal of sorting students into high and low "abilities," as Gerald Bracey, a former professor at George Mason University noted:

In Japan there is strong consensus that children should not be subjected to measuring of capabilities or aptitudes and subsequent remediation or acceleration during the nine years of compulsory education. In addition to seeing the practice as inherently unequal, Japanese parents and teachers worried that ability grouping would have a strong negative impact on children's self-image, socialization patterns, and academic competition.[3]

Lisa Yiu, a student of mine at Stanford, was interested in the very different ways Japan and the United States treat the grouping of students. She visited Japan and interviewed some of the mathematics teachers she observed, who explained to her why they do not use ability grouping. "In Japan what is important is balance. Everyone can do everything; we think that is a good thing. . . . So we can't divide by ability."

Japanese education emphasizes group education, not individual education. Because we want everyone to improve, promote, and achieve goals together, rather than individually. That's why we want students to help each other, to learn from each other . . . , to get along and grow together—mentally, physically and intellectually.[4]

The Japanese approach of teaching students "to help each other, to learn from each other, . . . to get along and grow together—mentally, physically and intellectually" is part of the reason for their high achievement. Research tells us that approaches that keep students as equal as possible and that do not group by "ability" help not only those who would otherwise be placed in low tracks, which seems obvious, but also those who would be placed in high tracks too.

In neither of my two longitudinal studies of students working in different mathematics approaches did I intend to research the impact of ability grouping, but it emerged as a powerful factor in both of them. In both studies the two successful schools were those that chose against ability grouping and used a different system of grouping, a result that mirrors the larger-scale international findings. England uses an overt system of ability grouping, one much harsher in its impact on low achievers than any system in the United States. In England, students are placed into "sets" that are ranked by achievement. I studied students in sets across the range from set 1 (the highest) to set 8 (the lowest). It did not surprise many people that students who were put into low sets achieved at low levels, partly because of the low-level work they were given and partly because students gave up when they knew they had been put into a low set and labeled a low achiever. This was true of students from set 2 downward. What was more surprising to people was that many of the students in set 1, the highest group, were also disadvantaged by the grouping. Students in set 1 reported that they felt too much pressure from being in the top set. They felt the classes were too fast, they were unable to admit to not following or understanding, and many of them started to dread and hate math lessons. The students in the highest group should have felt good about their understanding of math, but instead they felt pressured and inadequate. After three years of ability-grouped classes, the students achieved at significantly lower levels than students who had been grouped in mixed-ability classes.

In the United States the ability grouping that is used is nowhere near as overt as that in England, but it still has a significant impact. At around seventh and eighth grade, students in the United States typically get placed into different levels of classes, which determines their future for many years to come. Despite the importance of the different placements, the names of the classes often sound innocuously similar. Some seventh-grade students may be in something called "math 7" while their peers are in pre-algebra, a higher class. Or eighth graders may be placed in "math 8" or in pre-algebra, while their peers are in algebra. The critical information that schools rarely provide is that in most American high schools, students cannot take calculus unless they have already passed algebra in middle school. If, like in most high schools, classes are a year long, then students who take algebra in ninth grade will take four years of courses without ever reaching calculus (algebra, geometry, advanced algebra, and precalculus). Thus the tracking decisions made by middle school teachers impact the classes reached in high school and, from there, students' chances of being admitted to colleges of their choice. Middle school students should hear a strange sound when they are placed into lower-level math classes. It is the sound of doors closing.

As educators become more aware of the disadvantages of tracking, more schools are trying different approaches. Carol Burris, from South Side High School in New York, with Columbia University professors Jay Heubert and Hank Levin, conducted a study of a detracking innovation in mathematics.[5] They compared the performance of students who were in tracked classes with those who were in mixed-ability classes. The researchers compared six annual cohorts of students in a middle school in a district of New York. The students attending the school in the first three years were taught in tracked classes with only high-track students being taught the advanced curriculum. But in the next three years all students in grades six through eight were taught the advanced curriculum in mixed-ability classes and all of the eighth graders were taught an accelerated algebra course. The researchers looked at the impact of these different middle school experiences upon the students' completion of high school courses and their achievement, using four achievement measures, including scores on the Advanced Placement calculus examinations. They found that the students from detracked classes took more advanced classes, pass rates were significantly higher, and students passed exams a year earlier than the average in New York State. Their scores were also significantly higher on various achievement tests. The increased success from detracking applied to students across the achievement range—from the lowest to the highest achievers.

In my own work in the United States and England, I have studied high schools that have shown a commitment to more equal schooling by placing all incoming students into mixed-ability classes. Railside High School in California started all students in a class called algebra regardless of prior achievement. This was a challenging class for all students, even those who had taken algebra in middle school, because it involved working on complex, multidimensional, multilevel problems. The school also taught classes that were ninety minutes long but that only took half a year. This meant that all students could reach calculus as they had eight opportunities to take a math class during their four years of high school. The results were outstanding, including 47 percent of the seniors taking calculus and precalculus advanced classes, compared to 28 percent of students in the more typical tracked high schools. Interestingly, the students who were most advantaged by the mixed-ability grouping were the highest-achieving students, who achieved at higher levels than students who were placed into high tracks in the other schools, and who improved their achievement more than any other students in their school.

Many parents fear mixed-ability grouping and cannot see the logic of grouping students with very different needs and limited teacher resources into one class. So why is it that mixed-ability grouping is repeatedly found to be associated with higher achievement? The most important reasons are explained next.

*high teacher
expectations*

Opportunity to Learn

Researchers consistently find that the most important factor in school success is what they call "opportunity to learn."[6] If students are not given opportunities to learn challenging and high-level work, then they do not achieve at high levels. We know that when students are in lower groups, they receive low-level work and this, in itself, is damaging. In addition, teachers inevitably have lower expectations for students. In the 1960s Robert Rosenthal and Lenore Jacobson, two sociologists, conducted an experiment to look at the impact of teacher expectations. Students in a San Francisco elementary school were assigned to two groups. Both groups were taught the same work, but teachers were told that one group included all the students identified as especially talented. In reality, the students had been randomly assigned to the two groups. After the experimental period, the students in the group identified as talented showed better results and scored at higher levels on IQ tests. The authors concluded that this effect was entirely due to the different expectations the teachers held for the students.[7] In a study of six schools in England, a team of researchers and I were saddened to find that teachers routinely underestimated children in low groups. One student reported to us: "Sir treats us like we're babies, puts us down, makes us copy stuff off the board, puts up all the answers like we don't know anything. And we're not going to learn from that, 'cause we've got to think for ourselves."

The students talked openly to us about the low expectations teachers held for them and the way their achievement was being held back, and observations of the classes confirmed this to be true. The students simply wanted to be given opportunities to learn: "Obviously we're not the cleverest, we're group five, but still—it's still maths, we're still in year nine, we've still got to learn."[8]

When teachers have lower expectations for students and they teach them low-level work, the children's achievement is suppressed. This is the reason that ability grouping is illegal in many countries in the world, including Finland, a country that tops the world in the international achievement tests.[9]

Tracking Leads to Damaging Fixed Mind-sets

When students are put into ability groups, they get hit with a damaging message—that their "ability" is fixed. This leads many students with healthy growth mind-sets—that is, they believe that the harder they work the smarter they get—to change to a fixed mind-set, believing that they are either smart or that they are not. Fixed mind-sets, as explained in the preface, result in significantly lower achievement for students and disadvantages that continue throughout life. In one of Carol Dweck's then doctoral students' studies, Carissa Romero found that when students moved into ability groups in grade seven, positive growth-mind-set thinking reduced, and the students who were most negatively affected were those going into the top track. Many more of the students started to believe that they were smart, and this sets students on a very vulnerable pathway, where they become fearful of making mistakes and start to avoid more challenging work. The consequences, especially for high-achieving girls, are devastating.

Student Differences

When students are placed into a tracked group, high or low, assumptions are made about their potential achievement. Teachers tend to pitch their teaching to students in the middle of the group, and they teach a particular level of content, assuming that all students are more or less the same. In such a system, the work inevitably is at the wrong pace or the wrong level for many students within a group. The lower students struggle to catch up, while others are held back. While a teacher of any class, including a mixed-ability class, can wrongly assume that the students all have the same needs, tracking is based on this erroneous assumption, and when teachers have a tracked group, they often feel license to treat all students the same. This is true even when students clearly have different needs and would benefit from working at different paces.

In a mixed-ability group, the teacher has to *open* the work, making it suitable for students working at different levels and different speeds. Instead of prejudging the achievement of students and delivering work at a *particular* level, the teacher has to provide work that is multileveled and that enables students to work at the highest levels they can reach. This means that work can be at the right level and pace for all students.

Borderline Casualties

When teachers assign students to different groups, they make decisions that affect their long-term achievement and their life chances. Despite the importance of such decisions, they are often made on the basis of insufficient evidence. In many cases, students are assigned to groups on the basis of a single test score with some students missing the high groups because of one point. That one missed point, which students may have scored on another day, ends up limiting their achievement for the rest of their lives.

Researchers in Israel and the UK found that students on the borders of different groups had essentially the same understandings, but the ones entering higher groups ended up scoring at significantly higher levels at the end of school because of the group they were placed into. Indeed, the group that students were put into was more important for their eventual achievement than the school they attended.[10] Borderline casualties are those students who just miss the high groups and become casualties of the grouping system from that day on. There are many of these students in schools and for them the almost arbitrary assignment to a lower group effectively ends their chances of success.

Student Resources

In a tracked class the main sources of help are the teacher and the textbook. Students are presumed to have the same needs and to work in the same way, so the teacher feels comfortable lecturing to the class for longer periods of time and requiring the class to work in silence or in very quiet conditions, denying them the many advantages of talking through problems, as set out in chapter 2.

In mixed-ability classes the students are organized to work with each other and help each other. Instead of one person serving as the resource to thirty or more students, there are many. The students who do not understand as readily have access to many helpers. The students who do understand serve as helpers to classmates. This may seem like it is wasting the time of high achievers, but the reason these students end up achieving at higher levels in such classrooms is because the act of explaining work to others deepens understanding. As students explain to others, they uncover their own areas of weakness

Explaining Key to learning

and are able to remedy them and they strengthen what they know. In the two longitudinal studies I conducted, the high-achieving students told me that they learned more and more deeply from having to explain work to others. Some of the high achievers from Railside High School talked about their experiences in mixed-ability groups in the United States. Zane had said, "Everybody in there is at a different level. But what makes the class good is that everybody's at different levels, so everybody's constantly teaching each other and helping each other out."

Some of the higher achievers entered Railside thinking it unfair that they had to explain work to others, but they changed their minds within the first year as they realized that the act of explaining was helping them. Imelda reflected: "So maybe in ninth grade it's like 'Oh, my God. I don't feel like helping them. I just wanna get my work done. Why do we have to take a group test?' But once you get to AP Calc, you're like 'Ooh, I need a group test before I take a test.' So, like, the more math you take and the more you learn, you grow to appreciate, like, 'Oh, thank God I'm in a group!'"

The high achievers also learned that different students could add more than they thought to discussions. As Ana said, "It's good working in groups because everybody else in the group can learn with you, so if someone doesn't understand—like if I don't understand but the other person does understand—they can explain it to me, or vice versa, and I think it's cool."

Students who work together, supporting each other's learning, provide a tremendous resource for each other, maximizing learning opportunities at the same time as learning important principles of communication and support.

Student Respect

When we consider the role of ability grouping and the difference it makes in students' lives, there is another dimension besides achievement that it is critical to consider. For ability grouping not only limits opportunities, it influences the sorts of people our children will become. As students spend thousands of hours in their mathematics classrooms, they do not only learn about mathematics, they learn about ways of acting and ways of being.

Mathematics classrooms influence, to a high and regrettable degree, the confidence students have in their own intelligence. This is unfortunate both because math classrooms often treat children harshly but also because we know that there are many forms of intelligence, and math classrooms tend to value only one. In addition to the power that math classrooms have to build or crush children's confidence, they also influence to a large extent the ideas students develop about other people.

Through my own research I have found that students in tracked classes in American high schools not only developed ideas about their own potential, but they began to categorize others in unfortunate ways—as smart or dumb, quick or slow. Comments such as this came from students in tracked classes: "I don't want to feel like a retard. Like if someone asks me the most basic question and I can't do it, I don't want to feel dumb. And I can't stand stupid people either. Because that's one of the things that annoy me. Like stupid people. And I don't want to be a stupid person."

The students who had worked in mixed-ability classes at Railside High School did not talk in these ways and they developed impressive levels of respect for each other. Any observer to the classrooms could not fail to notice the respectful ways students interacted with one another, seeming not to notice the usual dividing lines of social class, ethnicity, gender, or "ability." The ethnic cliques that often form in multicultural schools did not form at Railside, and students talked about the ways their math classes had taught them to be respectful of different people and ideas. In learning to consider different approaches to math problems, students also learned to respect different ways of thinking more generally and the people making such contributions. Many of the students talked about the ways they had opened their minds and their ideas—for example:

> **Tanita:** You got everyone's perspective, 'cause like when you're debating a rule or a method, you get someone else's perspective of what they think instead of just going off your own thoughts. That's why it was good with, like, a lot of people.
>
> **Carol:** I liked it too. Most people opened up their ideas.

Undeniably, one of the goals of schools is to teach students subject knowledge and understanding, but schools also have a responsibility to teach students to be good citizens—to be people who are open-minded, thoughtful, and respectful of others who are different from themselves. The Carnegie Corporation of New York, based on a report from the Council for Adolescent Development, recommended that detracking occur in schools in order to create "environments that are caring, healthy, and democratic as well as academic."[11] The Coalition for Essential Schools also recommended that schools abandon tracking for both academic and moral reasons. Its former director Ted Sizer argued that "if we want citizens who take an active and thoughtful part in our democracy they must get trained for this in school—working together on equally challenging problems, and using every possible talent toward their solutions."[12]

Although it seems to make sense to place students into groups where they have similar needs, the negative consequences of tracking decisions—in terms of students' achievement and their moral development—are too strong to ignore. Teachers of mathematics are often inexperienced with different systems of grouping, but this is not a good reason for perpetuating a flawed approach. Teachers need to be supported in gearing their teaching toward high achievement for all. I have visited math classes that have recently become detracked and seen teachers using the same approach they have always used, to disastrous effect. These teachers gave out worksheets targeting small areas of content appropriate for only a few students in the class, leaving the rest floundering or bored.

For mixed-ability classes to work well, two critical conditions need to be met. First, the students must be given open work that can be accessed at different levels and taken to different levels. Teachers have to provide problems that people will find challenging in different ways, not small problems targeting a small, specific piece of content. These are also the most interesting problems in mathematics so they carry the additional advantage of being more engaging. See www.youcubed .org for many examples of these engaging tasks. I have seen such problems used to great effect in a number of classrooms. These sorts of multilevel problems are used in Japanese classrooms in order to promote high achievement for all, as Steve Olson, author of the bestselling book *Count Down,* reflects:

> In Japanese classrooms . . . teachers *want* their students to struggle with problems because they believe that's how students come to really understand mathematical concepts. Schools do not group students into different ability levels, because the differences between students are seen as a resource that can broaden the discussion of how to solve a problem. Not all students will learn the same thing from a lesson; those who are interested in and talented at math will achieve a different level of proficiency from their classmates. But each student will learn more by having to struggle with the problem than by being force-fed a simple, predigested procedure."[13]

Japanese students are not all expected to learn the same from each lesson, which is the unrealistic expectation for many

American classes; instead, they are given challenging problems and each student gets the most from them that he or she can.

In addition to open, multilevel problems, the second critical condition for mixed-ability classes to work is that students are taught to work respectfully with each other. I have observed many math classrooms where students are working in groups, but the students do not listen to each other. The teachers of such classrooms have given students good problems to work on together, and they have asked students to discuss the problems, but the students have not learned to work well together. This can result in chaotic classrooms with groups where only some of the students do the work, or, even worse, groups in which some students are ignored or ridiculed by other students because they are deemed to be low status. Teaching students to work respectfully requires careful and consistent building of good group behavior. Some teachers do this by highlighting the need for respect and hard work in groups, some teachers employ additional strategies such as "complex instruction" (an approach designed for use with mixed-ability groups), which is aimed at reducing status differences between students. Whatever the approach, when students learn to work well and respectfully together and their different strengths are seen as a resource rather than a point to ridicule, then children are helped by being able to achieve at high levels and society is helped through the development of respectful, caring young people.

Psychological Prisons

In my study of Amber Hill and Phoenix Park in England, I was able to follow students through a school that used ability grouping (Amber Hill) and one that did not (Phoenix Park). At Amber Hill the teachers also taught traditionally, whereas at Phoenix Park the teachers used complex, open problems. As I described in chapter 3, I was fortunate in being able to catch up with the students eight years after my initial study and talk to them about the impact of their school experiences on their jobs and lives. During that study I found that one of the most important differences between the students, perhaps not surprisingly, was the grouping approaches they experienced. At Amber Hill, where they used ability grouping, the adults talked about how the grouping had shaped their whole school experience and many of those from set 2 downward talked not only about the ways their achievement had been constrained by the grouping but also the ways they had been set up for low achievement in life. In a statistical comparison of the jobs that the ex-students were working in, I found that those who had experienced mixed-ability grouping, despite growing up in one of the poorest areas in the country, were now in more professional jobs than those who had experienced tracking.[14]

Interviews with some of the young adults gave meaning to these interesting differences. The students from Phoenix Park talked about the ways the school had excelled at finding and promoting the potential of different students; they said that teachers had regarded everyone as a high achiever. The young adults communicated a positive approach to work and life, describing the ways they used the problem-solving approaches they had been taught in school to get on in life. The young adults who had attended Amber Hill, which had put them into sets, told me that their ambitions were broken at school and their expectations lowered. One young man, Nikos, spoke passionately about the ability grouping experience:

> You're putting this psychological prison around them. . . . People don't know what they can do, or where the boundaries are, unless they're told at that kind of age.
>
> It kind of just breaks all their ambition. . . . It's quite sad that there's kids there that could potentially be very, very smart and benefit us in so many ways, but it's just kind of broken down from a young age. So that's why I

dislike the set system so much—because I think it almost formally labels kids as stupid.

The impact that ability grouping has upon students' lives—in and beyond school—is profound. Researchers in England revealed that 88 percent of children placed into ability groups at age four remain in the same groupings until they leave school.[15] This is one of the most chilling statistics I have ever read. The fact that children's futures are decided for them by the time they are placed into groups, at an early age, derides the work of schools and contravenes basic knowledge about child development and learning. Children develop at different rates, and they reveal different interests, strengths, and dispositions at various stages of their development. In the United States such decisions are usually made during middle school, but they still prejudge children's potential before they have had a chance to develop. One of the most important goals of schools is to provide stimulating environments for all children—environments in which children's interest can be piqued and nurtured, with teachers who are ready to recognize, cultivate, and develop the potential that students show at different times and in different areas. This can only be done through a flexible system of grouping that does not prejudge a child's achievement and that uses multi-leveled mathematics materials that individual students take to their own highest level. Only such an approach will enable America to become a fairer society in which all children are given the chance to be successful.

6 / Paying the Price for Sugar and Spice

How Girls and Women Are Kept Out of Math and Science

When I started my research at Amber Hill, Caroline was fourteen years old, eager to learn, and very accomplished. When the students had entered the school some three years earlier, they had all been given a math test. Caroline had earned the highest score in her year, but three years after coming to the school she was the lowest-achieving student in her class. How could it all have gone so badly wrong? When I met Caroline, she had just been placed into the top set—a group that was taught by Tim, the friendly and well-qualified math department chair. But Tim was a traditional teacher and he, like most math teachers, demonstrated methods on the board and then expected students to work through exercises practicing the techniques. Caroline sat at a table of girls, six of them in total. All of the girls were high achievers and they all wanted to do well in math. From the beginning, Caroline looked uncomfortable in class. She was an inquisitive and thoughtful girl, and whenever Tim explained methods to the class, she, like many girls I have taught and observed over the years, had questions: Why does that method work? Where does it come from? And how does that fit with the methods we learned yesterday? Caroline asked Tim these questions from time to time, but he would generally just re-explain the method, not really appreciating why she was asking. Caroline became less and less happy with math and after a while her achievement started to decline.

In one of their lessons the students were learning about the multiplication of binomial distributions. Tim had taught students to multiply binomial expressions such as

$$(x + 3)(x + 7)$$

by telling students that they:

1. Multiply the first terms (x times x)
2. Multiply the outer terms (x times 7)
3. Multiply the inner terms (3 times x) and then
4. Multiply the last terms (3 times 7) Then add all the terms together ($x^2 + 7x + 3x + 21 = x^2 + 10x + 21$)

Students are often taught to remember this sequence with the pneumonic FOIL (first, outer, inner, last). These are the sorts of procedures in math class that seem meaningless to students— they are hard to remember and easy to muddle up. I approached Caroline's group one day as she was sitting with her head in her hands. I asked her if she was all right, and she looked at me with an agonized expression. "Ugh. I hate this stuff," she said. "Can you tell me why it works like this? Why does it have to be in that order, with all of that adding?" She had tried asking Tim, who had told her that that is how the formula works and you just need to remember it.

I knelt beside Caroline's desk and asked her whether I could draw a diagram for her. I explained that we could think about the multiplication visually by thinking of the two expressions as sides of a rectangle. She sat up and watched as I drew a sketch:

	x	7
x	x^2	7x
3	3x	21

Soon all the girls at the table were watching. Before I had finished the drawing, Caroline said, "Oooh, I see it now," and the others made similar appreciative noises. I felt a bit bad about offering this drawing, as my role in the classroom was not to help students and certainly not to undermine Tim's teaching, but it was an opportunity that I took on that one occasion. The drawing was simple but it offered a lot—it allowed the girls to see *why* the method worked and that was important for them.

I observed Tim's class, and other classes at the school, many times over the three years. As I interviewed more and more of the boys and girls, I started to notice that the desire to know *why* was something that separated the girls from the boys. The girls were able to accept the methods that were shown to them and practice them, but they wanted to know *why* they worked, *where* they came from, and how they *connected* with other methods. Some of the boys were also curious about the ways methods were connected and how they worked, but they seemed willing to adapt to a teaching approach that did not offer them such insights. In interviews the girls would frequently say such things as "He'll write it on the board and you end up thinking, 'Well, how come this and this? How did you get that answer? Why did you do that?'"

Many of the boys, on the other hand, would tell me that they were happy as long as they were getting answers correct. The boys seemed to enjoy completing work at a fast pace and competing with other students, and they did not seem to need the same depth of understanding. John (year ten) spoke for many of the boys when he said, "I dunno, the only maths lessons you like are when you've really done a lot of work and you're proud of yourself because you've done so much work, you're so much ahead of everyone else."

In a questionnaire I gave to the whole of the year cohort, I asked the students to rank five ways of working in mathematics. Ninety-one percent of girls chose "understanding" as the most important aspect of learning mathematics, compared with only 65 percent of the boys, which was a statistically significant difference.[1] The rest of the boys said that the memorization of rules was the most important. The girls and boys also acted differently in lessons. During the hundred or so lessons that I observed, I would often see boys racing through their textbook questions, trying to work as quickly as possible and complete as many questions as they could. I would, just as frequently, observe girls looking lost and confused, struggling to understand their work, or giving up all together. In lessons I would often ask students to explain to me what they were doing. The vast majority of the time, the students would tell me the chapter title and, if I asked them questions like "Yes, but what are you actually *doing*?" they would tell me the number in the exercise; neither girls nor boys would be able to tell me why they were using methods or what they meant.

On the whole the boys were unconcerned by this as long as they were getting their questions right, as Neil told me in an interview: "Some of the stuff you do, it's just hard, and some of it's really easy and you can remember it every time. I mean, sometimes you try and race past the hard bits and get it mostly wrong, to go onto the easy bits that you like."

The girls would get questions right, but they wanted more. As Gill explained: "It's like you have to work it out and you get the right answers but you don't know what you did. You don't know how you got them, you know?"

At the end of the three years, the students took a national examination. In the top set (similar to honors class in the United States), which I had observed so closely, the girls achieved at significantly lower levels than the boys, which was a pattern that was repeated in different groups across the year cohort. Caroline, once the glittering star of the group, achieved the lowest grade of all. By the time she was finishing the course, she had decided that she was no good at math, despite the fact that she had been the highest achiever in the school. Caroline, and many of the other girls, had underachieved because they had not been given the opportunity to ask why methods worked, where they came from, and how they were connected. Their requests were not at all unreasonable—they wanted to locate the methods they were being shown within a broader sphere of understanding. Neither the boys nor the girls particularly liked the traditional mathematics lessons at Amber Hill—math was not a popular subject at the school—but the boys worked within the procedural approach they were given whereas many of the girls resisted it. When they could not get access to the depth of understanding they wanted, the girls started to turn away from the subject. National statistics tell us that girls now do very well in mathematics, achieving at equal or higher levels than boys. This high achievement, given the inequitable approach that most girls experience, is testament to their capability and impressive motivation to do well. But the high achievement of girls often masks a worrying reality—the approaches they experience make many girls uncomfortable and their lack of opportunity to inquire deeply is the reason that so few women take mathematics to high levels.

At Phoenix Park School—where students were taught through longer, more open problems and they were encouraged to ask why, when, and how—the girls and boys achieved equally, and both groups achieved at higher levels than the students at Amber Hill on a range of assessments, including national exams.

Some years later, I was sitting in a math class myself, having decided to take an applied statistics course. We had a wonderful teacher, a woman who explained why and how methods worked—almost all of the time. I remember one day sitting in class when our teacher showed the formula for standard deviation and then said, "By the way, would you like to know why it works?" Then a funny thing happened. The women in the class chorused yes and most of the men chorused no. The women joked with the men, asking, "What is wrong with you?" One of the men responded quickly: "Why do we need to know why? It is better just to learn it and move on." It was then that I realized we were playing out the same gender roles as the students I had been observing at Amber Hill.

In one of my first research studies at Stanford, I decided to learn more about the experiences of high-achieving students in American high school classes. I chose six schools and interviewed forty-eight boys and girls about their experiences in AP calculus classes. In four of the schools the teachers used traditional approaches, giving the students formulas to memorize without discussing why or how they worked. In the other two schools the teachers used the same textbooks, but they would always encourage discussions about the methods that students were using. I was not investigating or looking for gender differences, but I was struck again by the reflections of the girls in the traditional classes and their need to inquire deeply, as Kate at the Lewis school described:

> We knew *how* to do it. But we didn't know *why* we were doing it and we didn't know how we got around to doing it. Especially with limits, we knew what the answer was, but we didn't know *why* or *how* we went around doing it. We just plugged into it. And I think that's what I really struggled with—I can get the answer, I just don't understand *why*.

Again, many of the girls told me that they needed to know *why* and *how* methods worked, and they talked about their dislike of classes in which they were just asked to memorize formulas, as Kristina and Betsy at the Angering school described:

> **K:** I'm just not interested in, just, you give me a formula, I'm supposed to memorize the answer and apply it, and that's it.
> **JB:** Does math have to be like that?
> **B:** I've just kind of learned it that way. I don't know if there's any other way.
> **K:** At the point I am right now, that's all I know.

Kristina went on to tell me that her need to explore and to understand phenomena was due to being a young woman:

Math is more, like, concrete, it's so "It's that and that's it." Women are more, they want to explore stuff and that's life kind of, like, and I think that's why I like English and science. I'm more interested in, like, phenomena and nature and animals and I'm just not interested in, just, you give me a formula, I'm supposed to memorize the answer and apply it, and that's it.

It was unfortunate for Kristina that mathematics was not one of her school subjects that allowed her to "explore" or to consider phenomena, when it should have been.

David Sela, from the ministry of education in Israel, and Anat Zohar, from the Hebrew University in Jerusalem, conducted an extensive investigation of gender differences in the learning of physics. They took my notion of the quest for understanding that I had found to be prevalent among girls in math classes and considered whether it was also prevalent among girls in physics classes. They found, resoundingly, that it was. The researchers drew from a database of approximately four hundred high schools in Israel that offered advanced-placement physics classes. They sampled fifty students from the schools and interviewed twenty-five girls and twenty-five boys. They found that the girls in physics classes were exhibiting the same preferences that I had found in mathematics classes, resisting the requirement to memorize without understanding, saying that it was "driving them nuts." The girls talked about wanting to know why methods worked and how they were linked. The authors concluded that "although both girls and boys in the advanced-placement physics classes share a quest for understanding, girls strive for it much more urgently than boys, and seem to suffer academically more than boys do in a classroom culture that does not value it."[2]

Neither the female math students I interviewed nor the female physics students interviewed in Israel wanted an easier science or math. They did not need or want softer versions of the subjects. In fact, the versions of mathematics and science they wanted required considerable depth of thought. In both cases the girls wanted opportunities to inquire deeply, and they were averse to versions of the subjects that emphasized rote learning. This was true of boys and girls, but when girls were denied access to a deep, connected understanding, they turned away from the subject.

The differences that have been found between girls and boys in mathematics and physics classes do not suggest that all girls behave in one way and all boys in another. Indeed, Zohar and Sela found that one-third of the boys they interviewed also expressed strong preferences for a deep, connected understanding. But they, like I, found that girls consistently expressed such preferences in higher numbers and with more intensity. Such gender differences are interesting, and they may also hold the key to understanding women's low levels of participation in STEM (science, technology, engineering, and mathematics) subjects.

The idea that girls and women value a different type of knowing was famously proposed by Carol Gilligan, an internationally acclaimed psychologist and author. In Gilligan's book *In a Different Voice,* she claimed that women are likely to be "connected thinkers," preferring to use intuition, creativity, and personal experience when making moral judgments. Men, she proposed, are more likely to be "separate" thinkers, preferring to use logic, rigor, absolute truth, and rationality when making moral decisions.[3] Gilligan's work met a lot of resistance, but it also received support from women who identified with the thinking styles she described. Some years later a group of researchers developed Gilligan's distinctions further, claiming that men and women differ in their ways of knowing, more generally. Psychologists Mary Belenky, Blythe Clinchy, Nancy Goldberger, and Jill Tarule,[4] proposed stages of knowing, and again claimed that men tended to be separate thinkers and women connected thinkers. The authors did not have a lot of data to support their claims that women and men think differently, and they received considerable opposition, which is understandable given that they were suggesting fundamental distinctions in the way women and men come to *think* and *know*. When I reported my own findings, that girls were particularly disadvantaged by traditional instruction that did not give them access to knowing how and why, I also received resistance. Indeed, some of my colleagues challenged me, saying that it was not possible that girls would have different preferences from boys in such a cognitive domain. But there are many reasons why girls may develop different preferences and a stronger push for understanding, ranging from known brain differences to vastly different socialization processes. For me the question of why girls pursue a depth of understanding in greater numbers than boys do is less important than the question of how we can change mathematics environments so that all students can understand deeply, and so that girls and boys not only can achieve at equal levels but can also pursue STEM subjects in equal numbers.

Classes in which students discuss concepts, giving them access to a deep and connected understanding of math, are good for girls and for boys. Boys may be willing to work in isolation on abstract rules, but such approaches do not give many students, girls or boys, access to the understanding they need. In addition, high-level work in mathematics, science, and engineering is not about isolated, abstract-rule following, but about collaboration and connection making.

There are plenty of boys who value and need connections

and communication and who choose other subjects because mathematics does not offer these, just as there are girls who can happily work in isolation without mathematical connections. If mathematics teaching included opportunities for discussion of concepts, for depth of understanding, and for connecting between mathematical concepts, then it would be more equitable and good for both sexes, and it would give a more accurate depiction of mathematics as it is practiced in high-level courses and professions.

Where Are We Now?

Given the ways in which mathematics is commonly taught and the preferences of many girls for deep understanding and inquiry, it is perhaps surprising that girls do as well as they do in mathematics—and they do perform very well. In 2010, women made up 42 percent of math majors and 41 percent of those taking masters degrees in mathematics and statistics. These numbers do not show equality and are not ones we should be complacent about, but they may show higher proportions of women than many think. Psychologists Janet Hyde, Elizabeth Fennema, and Susan Lamon produced a meta-analysis[5] of studies that have investigated gender differences in achievement, combining more than one hundred studies involving three million subjects. Even in 1990, with such a vast database, they found very small differences between girls and boys, with a huge amount of overlap.[6] Hyde and her colleagues argued that gender differences were too small to be of any importance and that they have been overplayed in the media, which has helped to create stereotypes that are damaging.

In most examinations in the United States there are also no recorded gender differences in mathematics. The small performance differences that exist take place only on the SAT[7] and the AP examinations (in 2002 the average score of girls was 3.3 compared to boys' 3.5). In England, a country with a similar education system, girls used to achieve at lower levels than boys on examinations, but now they achieve at higher levels in all subjects. In fact, the results for girls and boys in England have shifted in interesting ways over time. In England, at age sixteen, almost all students take the General Certificate of Secondary Education (GCSE) examination in mathematics. As mathematics is compulsory until age sixteen, equal numbers of boys and girls take this examination. In the 1970s, boys passed the mathematics GCSE examination[8] in higher numbers and achieved more of the highest grades; in the 1990s, girls and boys passed the exam in equal numbers but boys achieved more of the highest grades. By the 2000s, girls were passing the exam at higher rates than boys and achieving more of the highest grades.[9] In England, girls now perform at equal or higher levels than boys in mathematics and physics, on the GCSE and beyond, and they now achieve more of the highest grades in the most demanding high-level examinations.

Girls are doing very well now in the United States, England, and many other countries, but their strong performance hides a worrying fact—most mathematics classrooms are not equitable environments and girls often do well despite inequitable teaching. This is the reason that girls often opt out of mathematics even when they are achieving at high levels and even though a mathematical or scientific career could be very good for them. In high-level courses and mathematical jobs, the statistics are quite alarming. In 2009, women made up only 31 percent of mathematics PhDs, and in 2005 only 18 percent of mathematics faculty were women. The low numbers of women working as researchers and scientists across Europe is one of the priority areas for the European Union. Fifty-two percent of higher education graduates across Europe are women, but only 25 percent of these women take science, engineering, or technology subjects. Girls do well in math and science because they are capable and conscientious, but many do so through endurance, and math classroom environments are far from being equitable. Indeed, it is the impoverished version of mathematics that is offered to students that turns many people, female and male, away from the subject.

Other Barriers

Of course, the lack of opportunity to inquire deeply is not the only barrier to girls and women in mathematics and science. Mathematics classrooms in schools are considerably less gender stereotyped than they were twenty years ago, when sexist images prevailed in textbooks and mathematics teachers were found to give boys more attention, reinforcement, and positive feedback,[10] but still girls in some classrooms experience stereotyped attitudes and behaviors, contributing to their low interest and participation in math. Some school mathematics and science classrooms are also highly competitive, which deters many young women.

In university math departments the situation is worse. Abbe Herzig, a professor at the State University of New York at Albany, has produced evidence of the ways in which the climates of university mathematics departments can be icy cold toward women and minority students.[11] Herzig notes that women face many issues including sexism, stereotyped ideas about women's capabilities, feelings of isolation, and lack of role models,[12] especially at the graduate level. By and large, math departments in the United States remain a male preserve where the underrepresentation of women among students is eclipsed only by the underrepresentation of women among the faculty. There could not be a clearer statement about the absence of women in the history of the department or the lack of concern for their sense of inclusion now.

The teaching in mathematics departments can also be highly rule-based, which again denies girls the opportunity to ask why and how, as Julie, one of the young women who had given up on her mathematics degree at Cambridge University, explained to me:

> I think it was my fault because I did want to understand every single step and I kind of wouldn't think about the final step if I hadn't understood an in-between step. . . . I couldn't really see *why* they, *how* they got to it. Sometimes you want to know; I actually wanted to know.

For Julie, who had won awards for her mathematics achievement prior to attending Cambridge, her desire to understand how and why methods worked stopped her from going forward with the subject.

Societal Stereotypes

In addition to the problems in university mathematics departments and schools, young women and men suffer from the stereotypes that are perpetuated in society, particularly by the media. Ideas such as girls being too nurturing and caring to work in the hard sciences are based on incorrect ideas about women and about the ways mathematics and sciences work. Girls and women do not need a softer version of mathematics. Indeed, the sorts of inquiries women need could be said to epitomize true mathematical work, with its need for proof and rigorous analysis of ideas.

When girls went ahead of boys in mathematics and science (and all other subjects) in England, alarm bells rang everywhere.

Suddenly there was government money to look into gender relations in math and science, something that had never happened when girls were underachieving. It was interesting that, whereas people had always decided that the underachievement of girls in math and science was due to their intellect, when it was boys who were underachieving, people looked to external reasons—with suggestions that the books must be biased, the teaching approaches favored girls, or that teachers must be encouraging girls more. Nobody suggested that boys did not have the brains for math or science. Michele Cohen, a historian, gives an interesting perspective on the tendency of people to locate girls' underachievement *within* girls. She points out that this has been done throughout history and that in the seventeenth century scholars went to enormous lengths to explain away the achievement of girls and the working classes, as it was boys, specifically upper-class boys, who were believed to possess true intellect. People at that time explained that the superior verbal competence noted among girls was a sign of weakness and that the English gentleman's reticent tongue and inarticulateness was evidence of the depth and strength of his mind. Conversely, women's conversational skills became evidence of the shallowness and weakness of their mind.[13] In 1787, the Reverend John Bennett, from the Church of England, argued that boys appeared slow and dull because they were thoughtful and deep and because "gold sparkles less than tinsel."[14]

The tendency to locate sources of underachievement within girls and to construct ideas about female inadequacy is also characteristic of much of the psychological research on gender. In my interviews with high school students, I frequently encounter stereotypes about the potential of girls and boys. But it is particularly disturbing when I find that ideas about girls being mathematically inferior have come from the reporting of research. In a recent interview with a group of high school students in California, I asked Kristina and Betsy about gender differences:

> **JB:** Do you think math is different for boys and girls or the same?
>
> **K:** Well, it's proved that boys are better in math than girls, but in this class, I don't know.
>
> **JB:** Mmm, where do you hear that boys are better than girls?
>
> **K:** That's everywhere—that guys are better in math and girls are, like, better in English.
>
> **JB:** Really?
>
> **B:** Yeah. I watched it on *20/20* [a television current affairs program] saying girls are no good, and I thought, "Well, if we're not good at it, then why are you making me learn it?"

The girls refer to a television program that presented the results of research on the differences between the mathematical performance of girls and boys. The problem with some research on equity is not that researchers noted that girls were achieving less than boys, or that girls displayed less confidence in math classrooms, but that such findings were often presented as being due to the nature of girls rather than any external sources. This led educators to propose interventions that were well intentioned but that aimed to change girls. The 1980s spawned numerous programs that were intended to make girls more confident and challenging.[15] The idea behind such programs is often good, but they also lay the responsibility for change at the feet of the girls rather than mathematics teaching environments or the broader social system. On July 5, 1989, the *New York Times* ran the headline "Numbers Don't Lie: Men Do Better than Women" with an article discussing gender differences in SAT scores. But this article, like many others, used performance differences to suggest that women were mathematically inferior, rather than questioning the teaching and learning environments—as well as the biases they themselves were helping to create—that caused the underachievement of women. Now that women are ahead in most areas, it is interesting to note the absence of any analogous headlines proclaiming women's innate superiority.

It was once suggested that girls are made of "sugar and spice and all things nice"—a harmless nursery rhyme perhaps, but the idea that girls are sugary sweet and lack the intellectual rigor for mathematics and science is still around. It is time that such ideas are buried and that girls are encouraged to go into math and science, for their sake as well as the sake of the disciplines themselves. Mathematics is and has always been about deep inquiry, connection making, and rigorous thought. Girls are ideally suited to the study of high-level mathematics—and the only reason that they drop math in high numbers is because the subject is misrepresented and taught badly in too many classrooms in America.

I was invited to the White House recently to present to the Commission for Women and Girls on the reasons too few women choose STEM subjects. That day I told the gathered group that mathematics is one of the reasons girls do not go forward in STEM because girls look for a depth of understanding that is often unavailable in math classrooms. I also urged the group to pay more attention to teaching (instead of only focusing on stereotypes, messages, and role models) because I believe this is neglected in policy discussions. I look at initiatives to involve girls, such as in summer camps and after-school clubs, and I am aware that they have a positive impact, but they are working by trying to *change the girls* rather than by changing the teaching environments that push girls away from STEM. We urgently need to reorient mathematics and other subjects so that they focus on understanding and deep inquiry. When such changes are made girls choose STEM subjects in equal numbers to boys, and this is a goal that we should prioritize in the United States, not only for the futures of girls and women but for the future of the STEM disciplines.

Angela Wayne

2, 3, 5, 7, 11, 13, 17, 19, 23, 29, 31, 37, 41, 43, 47, 53, 59, 61, 67, 71, 73, 79, 83, 89, 97

7 / Key Strategies and
Ways of Working

In my work with the PISA team in Paris we have found that students take very different approaches to math and that the different approaches they take influence greatly whether they are successful or not. The incredible data set from PISA, including thirteen million students worldwide, shows that students who use memorization as their primary strategy are the lowest-achieving students in the world. This is consistent with my analyses of mathematics teaching and learning over the decades. I think of students as setting out on two contrasting pathways with mathematics: one of them leads to success and enjoyment and the other leads to frustration and failure. Students start on these pathways from a very early age. In this chapter I will review a study that illustrates the different pathways beautifully, as well as their relationship with achievement, before describing the time that my graduate students and I set out to change students' pathways through a summer school teaching program. What we achieved in a five-week period, in a challenging setting, is eminently achievable by teachers over the course of a year, and it is also achievable by parents armed with the right knowledge from research and the ideas I will share. I also include the stories of some children who changed as a result of the summer school, as their different circumstances and reasons for underachieving in mathematics may remind readers of their own children and students and the barriers they often face.

A Critical Way of Working

In an influential research study, two British researchers, Eddie Gray and David Tall, identified the reasons why many children struggle with math.[1] The results were so important that they should be shouted from the rooftops and posted in every math classroom across America.

Gray and Tall conducted a study of seventy-two students between the ages of seven and thirteen. They asked teachers in England to identify children from their classes whom they regarded as above average, average, or below average, and they interviewed the seventy-two children. The researchers gave children various addition and subtraction problems to do. One type of problem required adding a single-digit number, such as 4, to a teen-digit number such as 13. They then recorded the different strategies children used. These strategies turned out to be critical in predicting children's achievement.

For example, let's take 4 + 13.

$$
\begin{array}{c}
\bigcirc\ \bigcirc\bigcirc\bigcirc \\
+ \\
\bigcirc\ \bigcirc\bigcirc\bigcirc\bigcirc\ \bigcirc\bigcirc\bigcirc \\
\bigcirc\ \bigcirc\bigcirc\bigcirc\bigcirc
\end{array}
$$

One strategy for solving this addition problem is called "counting all." With this strategy students look at the 4 dots and count them (1-2-3-4); they then look at 13 dots and count them (1-2-3-4-5-6-7-8-9-10-11-12-13); they then look at all the dots and count them, from 1 to 17. This is often the first strategy that children use as they are learning to count.

A more sophisticated strategy that develops from "counting all" is called "counting on." A student using this strategy would count from 1 to 4 and then continue from 5 to 17.

A third strategy is called "known facts"—some people just know, without adding or thinking, that 4 and 13 is 17 because they remember these number facts.

The fourth strategy is called "derived facts." This is where students decompose and recompose the numbers to make them more familiar numbers for adding and subtracting. So they may say, "Well, I know 10 and 4 is 14," and then they add on the 3.

This sort of strategy of decomposing and recomposing numbers is helpful when you are given calculations to do, especially when doing them in your head. For example, you may need to know the answer to 96 + 17. To most people that is a nasty addition sum that looks daunting, but if 4 is taken from the 17 first and added to the 96, the problem becomes 100 + 13, which is much more reasonable. People who are good at mathematics decompose and recompose numbers all the time. This is the strategy that the researchers called "derived facts" as the students changed the numbers into ones that they knew the answers to, by decomposing and then recomposing the numbers. A better, well-known term for this flexible use of numbers is "number sense." Students with number sense are able to use numbers flexibly, grouping numbers, decomposing and recomposing.

The researchers found that the above-average children in the 8+ age group counted on in 9 percent of the cases, they used known facts 30 percent of the time, and they used number sense 61 percent of the time. In the same age group the students who were below average counted all 22 percent of the time, counted on 72 percent of the time, used known facts 6 percent of the time, but they never used number sense. It was this absence of number sense that was critical to their low achievement.

When the researchers looked at ten-year-olds, they found that the below-average group used the same number of known facts as the above-average eight-year-olds, so you could think of them as having learned more facts over the years but, noticeably, they were still not using number sense. Instead, they were counting. What we learn from this, and from other research, is that the high-achieving students don't just know more, but they work in very different ways—and, critically, they engage in flexible thinking when they work with numbers, decomposing and recomposing numbers.

The researchers drew two important conclusions from their findings. One was that low achievers are often thought of as *slow* learners, when in fact they are not learning the same things slowly. Rather, they are learning a *different* mathematics. The second is that the mathematics that low achievers are learning is a more difficult subject.

As an example of the very difficult mathematics that the below-average children were using, consider the strategy of counting back, which they frequently used with subtraction problems. For example, when they were given problems such as 16 – 13, they would start at the number 16 and count down 13 numbers (16-15-14-13-12-11-10-9-8-7-6-5-4-3). The cognitive complexity of this task is enormous and the room for mistakes is huge. The above-average children did not do this. They said, "16 take away 10 is 6, and 6 take away 3 is 3," which is much easier. The research showed that the students who were achieving at high levels were those who had worked out that numbers can be flexibly broken apart and put together again. The problem for the low-achieving children was simply that they had not learned to do this. The researchers also found that when low achievers failed at their methods, they did not change their method; instead, they fell back into counting more and more. Indeed, many of the low achievers became very efficient with small numbers, which lulled them into a sense of security. The low-achieving students came to believe that in order to be successful, they needed to count very precisely. Unfortunately, problems become more and more difficult in mathematics, so as time went on, the low achievers were trying to count in more and more complex situations. Meanwhile, the high achievers had forgotten counting strategies and were working with numbers flexibly. This is an easier task, but it is also a more important way of working in mathematics. As the low achievers continued to count, the high achievers worked flexibly and pulled further and further ahead.

Not surprisingly, the researchers found that the lower-achieving students who were not using numbers flexibly were also missing out on other important mathematical activities. For example, one of the important things that people do as they learn mathematics is compress ideas. What this means is, when we are learning a new area of math, such as multiplication, we may initially struggle with the methods and the ideas and have to practice and use it in different ways, but at some point things become clearer, at which time we compress what we know, and move on to harder ideas. At a later stage when we need to use multiplication, we can use it fairly automatically, without thinking about the process in depth.

William Thurston was a professor of mathematics and computer science at Cornell University who won the highest honor awarded in mathematics—the Fields Medal. He described the process of learning mathematics well:

Mathematics is amazingly compressible: you may struggle a long time, step-by-step, to work through the same
process or idea from several approaches. But once you
really understand it and have the mental perspective to
see it as a whole, there is often a tremendous mental
compression. You can file it away, recall it quickly and
completely when you need it, and use it as just one step
in some other mental process. The insight that goes with
this compression is one of the real joys of mathematics.[2]

One way of thinking about the learning of mathematics, visually, is to think of a triangle such as that shown in the following
diagram. The larger space at the top of the triangle is new mathematics you learn, that you need to think about and connect to
other areas, and it takes up a big space in your brain. The smaller
area at the bottom of the triangle represents mathematics that you
know well and has been compressed.

shutterstock/molars

It is this compression that makes it easy for people to use concepts they learned many years ago, such as addition or multiplication, without having to think about how they work every time they
use them. Gray and Tall found that the low-achieving students were
not compressing ideas. Instead, they were so focused on remem-
bering different methods, stacking one new method on top of the

next. Our brains can only compress concepts, not rules or methods, and the low-achieving students were not thinking conceptually, probably because they had been led to believe that mathematics is all about rules. For the low-achieving students the learning of mathematics would not be represented as a triangle. It would be more like a never-ending ladder stretching up to the sky, with every rung of the ladder being another rule or procedure to learn.

Students who are not taught to flexibly use numbers often cling to methods and procedures they are taught, believing that each method is equally important and must simply be remembered and reproduced carefully. For these students the mathematics they are learning is much more difficult.

Students who struggle to master more and more procedures, without using numbers flexibly or compressing concepts, are working with the wrong model of mathematics. These students need to work with someone who will change their pathway to mathematics and show them how to use numbers flexibly and how to think about mathematical concepts. But instead of working with people who will change the students' approach, what typically happens is that they get labeled as low-achieving students and people decide they need more drill, putting them into classes where they repeat methods over and over again. This is the last thing these students need and it simply feeds into their faulty worldview of math. Instead, they need opportunities to play with numbers, as I will suggest in the next chapter, and to develop "number sense." Fortunately, students can learn to use numbers flexibly and to consider concepts at any age, and it was with this knowledge that my graduate students and I set out to work with students who had previously been low achievers. We sought to give them the experience of using mathematics flexibly. This is the story of what happened.

The Summer of Math

As I prepared to teach students in sixth and seventh grades in California classrooms, I was a little nervous—it had been some years since I had taught mathematics in schools and all of my experience had been in London. The setting was a five-week summer school class in the San Francisco Bay area. Summer school classes are not known for their serious mathematics work. The students are there for a short time, so it is difficult to establish careful and good classroom routines. The classes were also extremely mixed, combining students who loved math and wanted to spend more time with the subject at the end of the regular school year, with those who had been forced to attend because they were failing in school. Reflecting this, 40 percent of the students had gained A's or B's in the previous year and 40 percent had gained D's or F's.

Four of my graduate students—Nick, Tesha, Emily, and Jennifer—and I taught four classes of sixth and seventh graders for two hours each day, four days a week. Our teaching, and the students' learning, was the focus of a research study—lessons were observed and students were given surveys, interviews, and assessments to monitor their learning. The classes were diverse both racially (39 percent Hispanic, 34 percent white, 11 percent

African American, 10 percent Asian, 5 percent Filipino, 1 percent Native American) and socioeconomically. I will start with a brief description of the summer school teaching and then illustrate the impact of the teaching by recalling the experiences of four very interesting young people.

One of the goals of our summer teaching was to give students opportunities to use mathematics flexibly and to learn to decompose and recompose numbers. We also wanted students to learn to ask mathematical questions, to explore patterns and relationships, and to think, generalize, and problem solve, as all of these are critical ways of working in mathematics, yet often neglected in school classrooms. As most of the students had spent the past years going through worksheets and rehearsing procedures in math classes, this was quite a change for them. We decided to focus the summer school on algebraic thinking, as we thought that would be most helpful to the students in future years, and to focus upon critical ways of working including asking questions, using mathematics flexibly, reasoning, and representing ideas. These different ways of working are all critical to success in mathematics but are often overlooked in classrooms.

It would be reasonable to assume that students, particularly those in sixth and seventh grades, know how to ask questions in a math class. Question asking is an important part of being a learner and one of the most useful things a student can do. But research has shown that student questions decline as students move through school and are surprisingly rare in classrooms.[3, 4] Indeed, such research suggests that as students progress through school, they learn *not* to ask questions but instead to keep silent, even when they don't understand. The act of question asking has been shown to increase mathematics achievement and improve attitudes among students, and we know that students who ask a lot of questions are usually the highest achievers.[5] But while many teachers regard student questions as valuable, they do not explicitly encourage questions. In our teaching of the summer school classes we chose to encourage student questions and to teach students the qualities of a good mathematics question. We started our sessions by telling students how much we valued questions. When the students asked good questions, we would post them onto large pieces of paper that we hung up around the room. We also gave students mathematical problems and encouraged them to extend the problems by posing their own questions. When the students were interviewed during and after the summer, many of them mentioned that they had learned that question asking was a useful strategy in math classes.

A second aspect of math learning that we encouraged was that of mathematical *reasoning*. Students learn to reason through being asked, for example, to justify their mathematical claims, explain why something makes sense, or defend their answers and methods to mathematical skeptics.[6, 7, 8, 9] Students who learn to reason about situations and determine whether they have been correctly answered learn that mathematics is a subject that they can make sense of rather than just a list of procedures to memorize. When we talked to students (individually or in groups) or to the whole class, we would always ask them to tell us why they thought an answer made sense and to justify it to their peers. By the end of the summer, students were doing this for themselves, pushing each other to explain and justify when they talked about their mathematical ideas.

In addition to question asking and reasoning, we also highlighted the importance of mathematical representations. Proficient problem solvers frequently use representations to solve problems and communicate results. For example, they may transform a problem given in numbers into a graph or diagram that illuminates different aspects of the problem. Or they may choose a particular representation to highlight something that helps a collaborator understand better. Although representation is a critical part of mathematical work and it is often the first thing that mathematicians do, it is rarely taught in classrooms. In our summer school teaching, we gave the students problems that were displayed in different ways, particularly visually, with manipulatives (such as cubes and beads) and diagrams, and we asked the students to produce representations as part of their work. In interviews with the students, some of them told us that they had never *seen* a mathematical idea before, and that the different representations they had seen and learned had been extremely powerful for them.

In all of our problems we encouraged and valued the flexible use of numbers. One of the best methods I know for teaching students how to use numbers flexibly is an approach called "number talks" devised by leading educator Ruth Parker.[10, 11] In number talks, the teacher asks students to work individually and without paper and pen or pencil. The teacher puts a calculation (usually an addition or multiplication problem) onto the board or overhead and asks students to work out the answer in their heads. An example of a problem we posed was 18 × 5. Teachers ask the students to signal privately when they have an answer, usually by showing one thumb, but not by raising their hand as this is too public and puts other students under pressure, also turning the activity into a speed contest. The teacher then collects all the different methods students have used. When we posed 18 × 5, four students shared their different methods for working it out.

Method 1	Method 2	Method 3	Method 4
18 + 2 = 20	10 × 5 = 50	15 × 5 = 75	5 × 18 = 10 × 9
20 × 5 = 100	8 × 5 = 40	3 × 5 = 15	10 × 9 = 90
5 × 2 = 10	50 + 40 = 90	75 + 15 = 90	
100 - 10 = 90			

These different methods all involve decomposing and re-composing numbers, changing the original calculation into other equivalent calculations that are easier. As students offered these examples, others saw a flexible use of numbers for the first time.

As the students took part in number talks, they realized that there was no pressure to finish quickly and that they could use any method they were comfortable with, and they began to enjoy them a lot. Some students learned, for the first time ever, to decompose and recompose numbers, as Gray and Tall recommended, as they saw others doing this and realized that it was extremely helpful. Many of the students mentioned the number talks as a highlight of the summer, as they particularly liked the challenge, not to mention the experience of sharing and seeing different mathematical methods. Although mathematics has a reputation for being a subject of single methods— with each problem requiring one standard method that must be remembered—nothing could be further from the truth. Part of the beauty of mathematical problems is that they can be seen and approached in different ways and, although many have one answer, they can be answered using different approaches. When we asked students about aspects of the teaching that had been helpful, learning about different methods was one of the most frequently cited aspects, second only to collaboration with classmates.

On the first day of summer school we gave students a blank journal to develop and record their mathematical thinking during the five-week course. We wanted the journals to be a space for students to play with ideas—as well as a safe place to communicate with us for those who were afraid or unwilling to share ideas publicly. We collected the journals frequently to look for mathematical thinking that we could help with and to give students feedback. Few of the students had ever been given the opportunity to write about math before, or to keep organized notes, which turned out to be very important for some of them.

In most of the lessons, students worked together in small groups or with partners. In some lessons, we allowed students to choose their seats and work groups; at other times, we chose seating assignments to help students work productively and to give them experience of working with a range of people and ideas. Our decision to allow students to sit where they wished at times was part of our general commitment to the promotion and encouragement of student choice, autonomy, and responsibility. We emphasized to students that they were in charge of their own learning, encouraging them to make changes in their behavior to improve their learning in class.

During the summer, we combined different tasks and ways of working, as variety is very important in mathematical work. There are many valuable ways to work in math classes—including through lectures, student discussions, and individual work—and there are many important types of tasks that students can work on, from long applied projects to short questions including contextual and abstract investigations. But none of these methods or tasks should be used exclusively, as there is benefit to students experiencing a range of ways of working, especially as they will need to work in different ways in their jobs and private lives. Among students who experience traditional math classes, one of the biggest complaints (and surely the most reasonable) is that the classes are always the same. The monotony causes disaffection; it also means that students only learn to work as they have in class—using procedures that have just been shown to them. During our classes we spent some time discussing ideas as a class and some time discussing ideas in groups. We often gave the students long problems to work through with others, and sometimes gave them shorter questions on worksheets to work through alone. We gave them tasks in which they explored patterns, similar to the algebraic work given at Phoenix Park and Railside, and we also gave tasks that were applied, such as a soccer World Cup activity. In that task students needed to work out which teams would play each other and how many different games there would be, as an introduction to combinatorics. Sometimes students were given tasks to work on for a set amount of time; sometimes we allowed students to choose the tasks they worked on and to choose the amount of time they spent on them. Students were also encouraged to use their own ideas in extending problems and choosing methods that made the most sense to them. In all of our activities we encouraged students to ask questions, to represent, reason, and generalize, and to share and think about different methods. We also spent a lot of time building students' confidence, praising them for their work and their thinking when it was good.

At the end of our classes we gave students the same algebra test that had been given them some months before in their regular classes. The students scored at significantly higher levels, even though we had not seen or taught to the content of the tests and our teaching had been significantly broader than the tests' content. The average score before the summer was 48 percent, at the end of the summer it was 63 percent, a significant increase. In surveys, 87 percent of students reported that the summer classes were more useful to them than their regular classes and 78 percent reported that they had enjoyed the classes a lot. In interviews, all of the students were extremely positive about the summer classes. Many had told interviewers that their regular classes were boring and frustrating, partly because they were made to work in silence. When reflecting on the summer class, the students talked about their enjoyment and learning, in particular from the collaboration with peers, from learning about multiple methods, from the different strategies they learned, and from the opportunities they received to think and reason.

In addition to the positive reports given in interviews and anonymous surveys, there was a huge improvement in the participation of the students in class during the summer. In the first class we asked students to complete a survey that asked them whose idea it had been to come to summer school and whether they wanted to be there. This revealed that 90 percent of students had been made to come and most of them said that they did not want to be there. The main reasons given by those who said they did not want to attend were that it would be boring, that they were losing their summer, that they would rather be socializing with friends, and that it would be unnecessary. The students' initial participation in our classes reflected their lack of enthusiasm. In the first session, many either were quietly withdrawn, sitting with heads on their arms or hiding under hoods, or they socialized with friends, chatting loudly and resisting our requests to work. We were thrilled that as the summer progressed, students' participation changed dramatically. After only a few days, students began to arrive at the door of class enthusiastic to start, they were taking the math problems very seriously, they were interested in mathematical questions; they participated in whole-class discussions, and generally their interest began to shift from social to mathematical concerns.

When the students returned to their middle school classes in the fall, we kept track of their achievement. Researchers visited the students' classes to observe the teaching approaches they were experiencing and their participation. These observations worried us as we saw students sitting in rows, in silence, working through short, narrow questions requiring single procedures for solutions. Sadly, the students were returning to the same math environments that most had told us they hated, and students received limited or no opportunities to use the learning approaches emphasized in the summer. This did not give us much hope that the math classes we had taught over the summer, in which students had been so engaged and excited to learn mathematics, would have a long-term impact. It is very difficult for students to return to the same environment in which they have done badly before and do better, even after a summer of enjoying mathematics and learning new ways of working. This is why the home is such an important place to encourage students to work in good ways. It was pleasing that our students' math grades did improve significantly in the fall, whereas the control group who attended the same summer school but did not attend our classes did not improve theirs significantly. Unfortunately, but not surprisingly, the students' renewed enthusiasm and increased achievement did not last, and their grades had fallen again by the following quarter.

Some of the students who attended our classes gained higher grades in the quarter following the summer *and continued to achieve higher grades in math*. For these young people, the relatively short math intervention had given them a new approach to math that they were able to continue to use. When we interviewed Lisa, one of the students who had maintained her improved achievement, and asked her how the summer was helping her in the regular school year, she said, "When I don't know how to solve a problem the way the teacher does it, I have other ways to solve it." Melissa, a student who had achieved an F before the summer and then achieved an A afterward, told interviewers that the most useful part of the summer had been learning strategies—in particular "to ask a question if you're not sure" and to look for patterns. She also said, "I used to hate math and I thought it was boring, but in this class math was a lot of fun." Melissa's enjoyment and her learning of strategies certainly had a big impact on her achievement. I will describe the ways in which the strategies that we taught the students during the summer can be encouraged in the home in the next chapter.

The research results on the whole group of students we taught were very positive, but the stories of individuals are probably most interesting and most insightful in understanding why many children are turned off by math. Let's look at four of the students:

Rebecca, Who Needed to Understand

Over the thirty or so years I have been a schoolteacher and researcher, I have met many students like Rebecca—and usually they are girls. Rebecca was conscientious, motivated, and high-achieving. Even though she achieved A+ grades in mathematics, she did not feel that she was good at it. Rebecca, like many other students, could follow the methods the teachers demonstrated in her regular classes and reproduce them perfectly, but she wanted to understand mathematics and she did not feel that the procedural presentations of mathematics she experienced gave her access to understanding. Rebecca described her previous math classes as always the same—the teacher would start with a warm-up and then give students worksheets to go through individually. There were no classroom discussions, and the questions they worked through were shallow and procedural. In interviews Rebecca and her friend Alice were asked if they regarded themselves as "math people." When Rebecca said no, Alice protested, saying that Rebecca had won an award in math. I asked Rebecca about this and why she did not regard herself as a math person. She said it was because "I can't remember things well and there is so much to remember."

When students tell me that there is a lot to remember in mathematics, I know that they are being taught badly and the subject is being misrepresented. I know also that they are being overwhelmed by the procedures teachers show and have come to believe that they must memorize them all, instead of understanding the concepts that link the procedures and render memorization unnecessary. Rebecca explained to me that you "have to remember" in math and that it was hard to remember because "you don't use it in life." All of the procedural lessons she experienced made Rebecca feel a failure even though she gained A+'s. Our math classes were very different as we taught students the strategies and ideas that connect mathematical procedures and give access to understanding. The notes that teachers had written on Rebecca in their recommendations for summer school told us that she would not, under any circumstances, talk in class as she was a "selective mute"—that is, she chose never to speak. But as Rebecca came to enjoy the math problems we presented and understand the mathematics we taught, she not only talked in class but even chose to go to the board and show her work to the whole class. We were surprised and thrilled to see Rebecca participate so publicly and knew that her participation spoke to the mathematical understanding she was gaining for the first time.

Rebecca told us that she most appreciated the summer classes because they allowed her to understand mathematics. Asked about the differences between the summer class and previous math classes, she said: "We stretch the problems a lot more. Before we would just get answers and not really stretch them. . . . Like patterns you can really look at it and find how it grows. . . . Stretching the problems helps you understand the problem more. In this class you don't stop when you get the answer, you keep going."

Rebecca was clear that working on longer problems that she was able to explore and extend gave her access to an understanding she had never had before. Some readers may worry that students who are high achievers cannot learn in the same classes as students who get D's and F's, but Rebecca did not give any indication, in class or in interviews, that she was held back in such a mixed setting. On the contrary, she reflected in her journal that the summer school class "has been more useful because we take the time to make sure everybody understands everything and we use different methods of learning."

When Rebecca was asked what she had learned, she talked about many aspects of the class. She said that she had learned to generalize algebraic patterns "instead of just staring at them," that she had learned to multiply double-digit numbers in her head (from the number talks), and that she had learned strategies such as organization and question asking. Also, in her words, she had learned "to think beyond the answer to the problem."

As Rebecca summarized her experience, "I enjoy this class more than any other math class I have had because we are learning math in a fun way, and I think we can learn as much or more math this way as in a textbook."

Rebecca continued to get A+ grades in the year after our summer class, but we hoped that she no longer believed mathematics to be a subject that had to be remembered without understanding or enjoyment.

Jorge, Who Needed a Chance and an Opportunity[12]

Jorge came to our math class with a history of D's and F's in math—and in school in general. He walked into math class on the first day with a huge smile on his face, joking with three of his friends. As on most days, he was dressed in baggy blue jeans and an Oakland Raiders cap that he only took off after much cajoling from his teacher. During that first lesson Jorge and his friends laughed and talked their way through the time, doing very little mathematics. Jorge was a social force in the classroom. Funny and charming, he could pull his friends away from work in an instant. Watching Jorge in class, it was easy to see why he wasn't doing well in school. Jorge had all the markings of a bad boy who would try to get away with doing as little as possible in math class and also keep other students from doing good work.

Observing Jorge during the later weeks of summer school, it was still possible to catch glimpses of the bad boy behavior that we saw on the first days. He still took a while to get started on tasks, still made jokes during discussions, and still tried to whisper something to throw off other students going to the board. However, he was also taking mathematics *extremely* seriously and, as he wrote in one of the surveys, he was working harder than he had *in any other math class*. During discussions, he listened to his classmates and sometimes volunteered to present his own thinking. Over time, he presented more often and with fewer jokes to deflect attention away from his work. In both the whole class and in groups, he increasingly talked with classmates seriously about mathematics, rather than always shifting conversations into the social territory with which he was more comfortable and in which he, an underperforming mathematics student, had more status. In a particularly striking example of Jorge's move to take math more seriously, he and two other students spent over an hour during one lesson grappling with the generalization for a challenging pattern. They kept their focus only on the problem for almost this entire time, even moving to a different area of the room when two girls started working on one side of their table. Jorge did not just follow his peers' conversation; he also kept the other boys on the task, asking questions when he was confused and volunteering ideas. He was deeply engaged in this task for a long period of time, a striking turnaround from the first day when he seemed to go out of his way to avoid serious mathematical work.

Jorge's comments in journals and interviews reveal aspects of the summer environment that may have contributed to his willingness to work more seriously in the class (and, importantly for such a "cool" kid, to be *seen* working more seriously). Notably, Jorge says that he worked harder in the summer school class than his regular classes, although he also describes the summer school classes as "funner." He explains that he works harder in summer school because "in our [regular] class they give us, like, easy problems and all that. And in this class you give us hard problems to figure out. You have to figure out the pattern and all that."

Asked what advice he would give math teachers to help them teach better, he says that he would tell them to "give harder problems." These comments suggested to us that he appreciated being taken seriously as a capable mathematics student.

When Jorge discussed the hard problems he was given in the summer school class, he talked both about being able to "figure out" the solutions himself and about being given the time to do that. He explained that he enjoyed working on pattern problems because "We stay on it longer . . . so we can really get to know how to do pattern blocks and everything, and try to figure out the pattern."

Jorge also talked about the value of working in groups. Another piece of advice he said he would give to teachers was to put students in groups, because "You learn more from other people's ideas."

On the days that Jorge became deeply engaged in mathematics, he was working on problems that could be viewed in many different ways; and he was working with high-achieving boys whom he respected. Jorge left our classes proud of his math work; importantly, Jorge had experienced working with high-achieving students on challenging problems, and that had changed everything for him. He felt respected, and he could offer his own thoughts and ideas. For a student who probably had been placed into low tracks and given easy work in the past, the experience of working on hard problems and discussing ideas with others was transformative.

Alonzo, Who Needed to Use His Ideas[13]

Alonzo was a very popular student. He was always surrounded by other students, before and after class. Alonzo could be described as the strong, silent type because of his striking, tall, athletic build and quiet, fiercely observant ways. For the first few days, Alonzo would slip into class undetected and pull the brim of his baseball cap down, as if hiding, silently watching the activities unfold before him. As the summer progressed, Alonzo's behavior changed and we came to realize that the mathematics he was working on was allowing his curiosity and creativity to take root and bloom. As Alonzo experienced the opportunity to use his ideas in math, he started to become a different kind of student in class.

Like many other students who attended summer school, Alonzo had gained an F in his previous math class and was forced to enroll by his math teacher. Alonzo described his previous class as one where the teacher talked and the students plodded through worksheets, and where group-based collaborations were extremely rare. Math learning in that class was largely a repetitive, worksheet-based experience. As Alonzo described it: "In regular math class we had to go to work. We couldn't talk or we couldn't, like, [offer] 'Oh, can I help?' No. He would just give us a paper, a pencil and put us to work." In informal conversations during class, Alonzo described his previous class as boring and frustrating.

One activity that piqued Alonzo's curiosity and creativity was "Staircases." In this task students were asked to determine the total number of blocks in a staircase that grew incrementally from 1 block high, to 2 blocks high, to 3 blocks high, and so on, as a move toward predicting a 10-block–high staircase, a 100-block–high staircase, and, finally, algebraically expressing the total blocks in any staircase. Students were provided with a box full of linking cubes to build the staircases if they wished.

A 4-block-tall staircase; total blocks = 4 + 3 + 2 + 1 = 10

Midway through the time allotted for the activity, Alonzo seemed to be playing with the linking cubes and not working on the problem. Drawing near, we saw that Alonzo had decided to modify his staircase so that it extended in four directions. Thus a 1-block-high staircase had a total of 5 blocks, a 2-block-high staircase had a total of 14 blocks, and so on.

From a distance we had thought Alonzo was off task, but he

Alonzo's staircase: 5 blocks   *5 + 9 = 14 blocks*   *5 + 9 + 13 = 27 blocks*

was using his creativity and curiosity to create a problem that was more diagrammatically and algebraically involved than the one that we had given to him. Eventually other students began looking in on Alonzo's work and, after convincing themselves they had mastered the original problem, they too tried to do the "Alonzo staircase" problem.

The teacher of Alonzo's class was so impressed with his innovative approach to the problem, as well as his growing interest in class and willingness to push himself to work at high levels, that she decided to phone home and tell his parents of his achievements. During this call, his mother described Alonzo as a junior engineer who had designed several small projects around the house—among them was a mechanical pulley system using dental floss and a stack of pennies. He used this in his bedroom to turn the light switch off without getting out of bed. Alonzo's mother told how he locked himself in his room while working on the project, coming out only to gather additional materials. In recounting the project to his mother, Alonzo said he'd had to experiment with different-size stacks of pennies until he figured out the exact weight it took to turn off the light switch without breaking the dental floss. Despite his creativity and apparent interest, she explained that she hadn't heard a single good word about Alonzo's mathematical work since he was in third grade. Alonzo's mother was very grateful for the call and asked if it would be possible for us to keep in touch with her after the summer to help lobby his teachers for more opportunities in which Alonzo could explore mathematics and use his creativity and resourcefulness. As we watched Alonzo work and reflected on the achievements his mother described, it was difficult to understand how he could ever achieve F grades in math classes.

Over the course of the five weeks, the open, problem-solving, group-based format of the summer program not only allowed Alonzo to play out his mathematical curiosity, but also encouraged him to take a more visible role in the classroom. During an activity called "Cowpens & Bullpens" students had to determine how many lengths of fencing were required to contain an increasing number of cows, given certain fencing parameters. At the close of the activity, student volunteers were asked to share their solutions and Alonzo was the first to volunteer. He strode to the front of the room and carefully diagrammed his fencing strategy on the overhead projector, documenting his work numerically and, eventually, algebraically.

In that moment the young man who once hid behind his baseball cap had all but disappeared and, in his place, we saw a seemingly more confident math student willing to share his ideas with the entire class. Alonzo was one of the highest scorers on the algebra test we administered at the end of the summer, with an impressive 80 percent, some 30 percentage points higher than when he took the test in the school year. But when Alonzo returned to his regular classes and was required to work on short questions in silence, he again withdrew and the grade he received for the class was, again, an F.

Tanya, Who Needed Discussion and Variety

Tanya described herself as a people person, which was easy to understand as she spent much of her time in math class talking to others. Tanya and her friend Ixchelle would talk and laugh as they worked, often whispering conspiratorially. Tanya did not seem to be aware that teachers were noticing her behavior and they would frequently ask her to be quiet. Tanya's talkativeness must have concerned many of her teachers. Indeed, her previous math teacher recommended that she attend summer school and among her comments wrote, "Tanya has a voice that carries easily." This not-so-subtle reference to Tanya's assertive personality implied that her exuberance needed controlling. But Tanya's own perspective on her social needs was interesting. In interviews Tanya described herself as needing an environment where students could work together because, as she explained: "Then I know if I'm doing wrong. If my way isn't the only way, if my way can be easily comprehended, or if it's a little too hard so I should try something different."

But Tanya's previous math classes did not allow collaborative work, as Tanya described to us: "For the past year, math was the hardest because you're not supposed to talk, you're not supposed to communicate. . . . [In other subjects] they let you talk. . . . In math class, [the teacher says] just, like, 'Okay, get to work, no talking, be quiet, shhhhh!' "

Tanya went on to describe the class and the teacher as dominated by silence: "[It is] a whole hour of silence. That's a good class to [the teacher]."

Our summer classes were very different and, for Tanya and many other students, the collaboration they experienced in class was critical to their engagement. Tanya spoke eloquently about the opportunities afforded by the discussions, and her reasons for valuing them had nothing to do with socializing or enjoying her time but with understanding mathematics. For example: "You can do it multiple ways. . . . The way our schools [in the regular year] did it is, like, very black and white, and the way the people do it here [in summer school] it's like very colorful, very bright. You have very different varieties you're looking at; you can look at it one way, turn your head, and all of a sudden you see a whole different picture."

Tanya enrolled in summer school because she was not doing well in her previous math class. In our class Tanya did extremely well and achieved one of the highest scores on the final algebra test. In interviews Tanya reported that she "learned more in five weeks than in almost a year." This statement reflected Tanya's appreciation of the learning opportunities she received.

One key feature of the program that Tanya seemed to benefit from was the varied nature of the mathematical tasks and the ways in which they encouraged multiple methods and approaches. Tanya especially appreciated the tasks that included manipulatives (blocks, tiles, linking cubes), the pattern work, whole-class activities such as the number talks, group and pair work, journal writing, and in-class presentations. In interviews she remarked: "I've enjoyed finding the patterns and stuff like that and groups . . . and it really helped me. The number talk really helped me to do stuff mentally."

Tanya seemed to revel in the plurality of ways she had to engage in math and described the tasks as more challenging, interesting, accessible, and fun: "It was much funner. You not only got to, like, just see the problem, you got to think it, you got different parts . . . you got to smell it, you got to eat it. And then after you finish the task that's given to you, you need to have another assignment to be, like, well, what if this changed, and you did this with that, instead of, you know, that? It just opens your mind and makes it harder, a new way of thinking."

Through interviews in the summer and fall, as well as in-class observations, Tanya revealed how the open, group-based format of summer school allowed her to express herself socially and develop mathematically. Tanya, like many of our summer school students, wanted desperately to appreciate math and to experience it on her own terms. She was, as she described herself, a "people person" who strove to see the color in math and yet found herself in situations that were monotonous and gray. Tanya's positive experiences in the summer had an impact on her, and she tried very hard to engage with her regular math class in the fall, receiving a very respectable B grade, but by the second quarter she had dropped to a D again. Tanya's low achievement, which did not fit her potential or enthusiasm from the summer, can probably be explained by the silent and monotonous nature of her school math class, as she aptly described: "I would say . . . the only way to describe summer school is very colorful and then this [regular school year] is just still, ugghhh, black and white. And you just wanna ask, 'Can I have a little bit of yellow?'" Tanya's request and that of so many other students—to experience a mathematics that is varied and interesting and colorful—is eminently reasonable.

In our summer school we gave students activities to work on and strategies to use that I will share in the next chapter. We taught the students that mathematics is a subject with many different methods and that numbers can be used flexibly. We encouraged the students to have confidence in their own mathematical abilities, something that no doubt contributed to their significantly improved performances in the next math classes they attended. It was unfortunate that the students went back to impoverished versions of math teaching and that we could not stay with them to encourage their new ways of working and thinking, but parents *can* stay with their children and keep steering them in the right mathematical directions. In the next chapter I will set out the sorts of activities and settings that will give children the best mathematical start in life and that may encourage students of all ages to enjoy and succeed at mathematics.

8 / Giving Children
the Best Mathematical Start

Activities and Advice for Parents

Biographies of mathematicians who have made great discoveries and inventions are fascinating to read, but I am always struck by the fact that so many were inspired not by school teaching but by interesting problems or puzzles that were given to them by family members in their homes. I was one of the people given the greatest mathematical start in life because my mum brought home puzzles and shapes for me to play with as a young child. It was many years later, at age sixteen, that I was also inspired by a great math teacher, who asked her students to talk about mathematics, which gave me access to a deeper understanding than I had ever met before. It is important not to underestimate the role of simple interactions in the home, and the role of puzzles, games, and patterns, in the mathematical development and inspiration of young people. Such problems and puzzles can be more important than all of the short questions that children work through in math classes. Sarah Flannery, the young woman who won the European Young Scientist of the Year award when she was sixteen, for inventing a "breathtaking" algorithm, reflects upon the fact that as a child she was given puzzles to work on at home, and how these were more important in her own mathematical development than the years of math that she was taught in classes. Mathematical puzzles and settings are the ideal way for parents— or teachers—to encourage their children into math. This chapter will outline some of the ways that these puzzles and key ways of working in math can be introduced to children with great effect.

Mathematical Settings

All children start life being excited by mathematics, and parents can become a wonderful resource for the encouragement of their thinking. Mathematical ideas that may seem obvious to us—such as the fact that you can count a set of objects, move them around, and then count them again and you get the same number—are fascinating to young children. If you give children of any ages a set of pattern blocks or Cuisenaire rods and just watch them, you will see them do all sorts of mathematical things, such as ordering the rods, putting them into shapes, and making repeating patterns. At these times parents need to be around to marvel with their children, to encourage their thinking, and to give them other challenges. One of the very best things that parents can do to develop their children's mathematical interest is to provide mathematical settings and to explore mathematical patterns and ideas with them.

There are many books filled with great mathematical problems for children to do, but it is my belief that the best sort of encouragement that can be given at home does not involve sitting children down and giving them extra math work, or even buying them mathematical books to work on. It is about providing settings in which children's own mathematical ideas and questions can emerge and in which children's mathematical thinking is validated and encouraged. Fortunately, mathematics is a subject that is ideally suited to the provision of interesting settings that can encourage this.

Colored beads of different shapes, and strings

One of my former Stanford doctoral students, Nick Fiori, taught a number of mathematics classes in which he provided students with different mathematical settings—some photographs of the settings are given here—and encouraged students to come up with their own mathematics questions within these settings. Students of different ages and backgrounds, including those who had suffered very negative experiences with math in the past, set about posing important questions. In some

Nuts, bolts, washers, and colored tape

cases these questions led to math problems that were completely new and had never been solved before. Fiori documented the various mathematical ways that students worked, and he made a very good case for the encouragement of similar problem posing in school classrooms, for at least some of the time.[1] I agree with him, but would

Cards from the game SET

Dice of various colors

Snap cubes of various colors

A square lattice grid with colored pegs

also encourage parents to provide such settings in the home for children of all ages.

Children's play with building and LEGO blocks in the early years has been identified as one of the key reasons for success in mathematics all through school.[2] Indeed, the fact that boys are usually encouraged to play with building blocks more than girls is the reason that differences often occur in spatial achievement among boys and girls, which impacts mathematics performance greatly. Any sorts of play with building blocks, interlocking cubes, or kits for making objects is fantastically helpful in the development of spatial reasoning, which is fundamental to mathematical understanding.

In addition to building blocks, other puzzles that encourage spatial awareness include jigsaw puzzles, tangrams, Rubik's Cubes, and anything else that involves moving objects around, fitting objects together, or rotating objects. Mathematical settings need not be sets of objects. They can be simple arrangements of patterns and numbers in the world around us. If you take a walk with your child, you will stumble upon all sorts of items that can be mathematically interesting, from house numbers to gate posts. The creative mind at work will see mathematical questions and discussion points everywhere—there is always something mathematical that can be brought into focus, if we only remember that that is what we should be doing.

Sticks of different unit lengths with eyelets and string

In her essay "The Having of Wonderful Ideas"[3] Eleanor Duckworth, a professor of education at Harvard, makes an extremely important point: the most valuable learning experiences children can have come from their own thoughts and ideas. In Duckworth's essay she recalls an interview she had with some seven-year-old children in which she asked them to put ten drinking straws, cut into different lengths, in order, from the smallest to the biggest. When Kevin walked into the room, he announced, "I know what I'm going to do," before Duckworth had explained the task. He then proceeded to order the straws on his own. Duckworth writes that Kevin didn't mean, "I know what you're going to ask me to do." Rather he meant: "I have a wonderful idea

Measuring cups with simple fraction volumes and a bowl of water

Pinecones of various shapes and sizes

about what to do with these straws. You'll be surprised by my wonderful idea." Duckworth describes Kevin working hard to order the straws and then being extremely pleased with himself when he managed it. For Kevin, the experience of ordering straws was much more worthwhile because he was working on his own idea, instead of following an instruction. Research on learning tells us that when children are working on their own ideas, their work is enriched with cognitive complexity and enhanced by greater motivation.[4, 5, 6] Duckworth proposes that having wonderful ideas is the "essence of intellectual development" and that the very best teaching that a parent or teacher can do is to provide settings in which children have their most wonderful ideas. All children start their lives motivated to come up with their own ideas—about mathematics and other things—and one of the most important things a parent can do is to nurture this motivation. This may take extra work in a subject such as mathematics, in which children are wrongly led to believe that all of the ideas already have been had and their job is simply to receive them,[7, 8] but this makes the task even more important.

Puzzles and Problems

In addition to the provision of interesting settings, another valuable way to encourage mathematical thinking is to give children interesting puzzles to work on. Sarah Flannery has written a fascinating book, *In Code: A Mathematical Journey,*[9] which describes her mathematical development. It is a very useful resource for parents who would like to encourage the best mathematical start in life. I don't think that parents need to be the mathematics professor that her father was in order to be successful; they just need the enthusiasm. Sarah Flannery talks about the way her mathematical development was encouraged by working on puzzles in her home. Although she and her siblings much preferred outdoor sports, her father would give them intriguing puzzles to think about in the evenings and these captured their young minds. Because her father was a mathematics professor and because she was so good at mathematics, people have often assumed that she was given extra math help at home, but Flannery shares something very important with her readers. She says:

> Strictly speaking, it is not true to say that I or my brothers don't get any help with math. We're not forced to take extra classes, or endure gruelling sessions at the kitchen table, but almost without our knowing we've been getting help since we were very young—out-of-the-ordinary help of a subtle and playful kind which I think has made us self-confident in problem solving. Ever since I can remember, my father has given us little problems and puzzles. I have often heard, and still hear, "Dad give us a puzzle." These puzzles challenged us and encouraged our curiosity, and many of them made math interesting and tangible. More fundamentally they taught us how to reason and think for ourselves. This is how puzzles have been far more beneficial to me than years of learning formulae and "proofs."[10]

Sarah Flannery gives examples of the sorts of math problems she worked on as a child that caused her to be so good at mathematics. Here are three of my favorites:

The Two-Jars Puzzle: Given a five-liter jar, a three-liter jar, and an unlimited supply of water, how do you measure out four liters exactly?

The Rabbit Puzzle: A rabbit falls into a dry well thirty meters deep. Since being at the bottom of a well was not her original plan, she decides to climb out. When she attempts to do so, she finds that after going up three meters (and this is the sad part), she slips back two. Frustrated, she stops where she is for that day and resumes her efforts the following morning—with the same result. How many days does it take her to get out of the well? Note: This question is assuming that the rabbit jumps up three meters and falls two meters each day.

The Buddhist Monk Puzzle: One morning, exactly at sunrise, a Buddhist monk leaves his temple and begins to climb a tall mountain. The narrow path, no more than a foot or two wide, spirals around the mountain to a glittering temple at the summit. The monk ascends the path at varying rates of speed, stopping many times along the way to rest and eat the dried fruit he carries with him. He reaches the temple shortly before sunset. After several days of fasting he begins his journey back along the same path, starting at sunrise and again walking at variable speeds with many pauses along the way, finally arriving at the lower temple just before sunset. Prove that there is a spot along the path that the monk will occupy on both trips at precisely the same time of day.[11]

Flannery talks about the ways these puzzles encouraged her mathematical mind because they taught her to *think* and *reason,* two of the most important mathematical acts. When children work on puzzles such as these, they are having to make sense of situations, they are using shapes and numbers to solve problems, and they are thinking logically, all of which are critical ways of working in mathematics. Flannery reports that she and her siblings would work on problems that her father had set for them each night over dinner. I like the sound of this ritual although I also appreciate that it is a difficult one to achieve in a busy household at the end of a tiring day. And puzzles do not have to be set by parents for children to work on. They can be something that children and parents work on together, each week, each month, or more occasionally. Whether a daily ritual or something less frequent, puzzles are incredibly useful in mathematical development, especially if children are encouraged to talk through their thinking and someone is there to encourage their logical reasoning. If children get in the habit of applying logic to problems and to persisting with problems until they solve them, they will learn extremely valuable lessons for learning and for life.

Some recommended books of mathematical puzzles are listed in appendix C.

Asking Questions

When exploring mathematical ideas with young people it is always good to ask lots of questions. Children often enjoy thinking through questions and it will help them develop mathematical ways of thinking. Good questions are those that give you access to your children's mathematical thoughts, as these will allow you to support their development. When I am called over to students who are stuck in math classes, I almost always start with "What do you think you should do?" Then, if I can persuade them to offer any ideas, I ask, "Why do you think that?" or "How did you get that?" Often children who have been taught traditionally will think they are doing something wrong at this point and quickly change their answer, but over time my students get used to the idea that I am interested in their thinking, and I will ask the same questions whether they are right or wrong. When students explain their thinking, I am able to help them move forward productively, at the same time helping them know that math is a subject that makes sense and that they can *reason* their way through mathematics problems.

Pat Kenschaft has written a useful book for parents, *Math Power: How to Help Your Child Love Math, Even If You Don't,* in which she quotes Swarthmore professor Heinrich Brinkmann. This particular professor was known on the Swarthmore campus for being able to find something right in what every student said. No matter how outrageous a student's contribution or question, he could respond: "Oh, I see what you are thinking. You're looking at it as if . . ."[12] This is a very important act in mathematics teaching because it is true that unless a child has taken a wild guess, then there will be some sense in what they are thinking—the role of the teacher is to find out what it is that makes sense and build from there. A parent in the home can do the sort of careful inquiring and guiding as they help children with mathematics that it is hard for teachers in a classroom of thirty or more children to do. If children give an answer and just hear that it is wrong, they are likely to be disheartened, but if they hear that their thinking is correct in some ways and they learn about the ways that it may be improved, they will gain confidence, which is critical to math success.

Children should also be encouraged to ask questions of themselves and others. I work with an inspirational teacher, Carlos Cabana, who, when asked for help, prompts students to pose a specific question that he and they can think about. As the students phrase their questions specifically, they are then able to see the mathematics more clearly, and they are often able to answer the questions themselves! Another great teacher I work with, Cathy Humphreys,[13] always says that she never asks a question that she knows the answer to. What she means by this is that she always asks students about their mathematical methods and reasons, which she could never know in advance. These are the most valuable questions for any math teacher to ask, as they give the teacher access to students' developing mathematical ideas. Pat Kenschaft puts it well: "If you can tap into the real thoughts of the person before you, you can untangle the knots around their mathematical inner light."[14] The ideal way in which to tap into learners' mathematical thoughts and to discover their mathematical inner light is to provide interesting settings and problems, to gently probe and question, and to encourage their thinking and reasoning.

When you are working with your child on math, be as enthusiastic as possible. This is hard if you have had bad mathematical experiences, but it is very important. Parents, especially mothers of girls, should never, ever say, "I was hopeless at math!" Research tells us that when mothers tell their daughters that they were not good at math in school, their daughter's achievement immediately goes down that same term.[15] It is important not to share your own negative feelings, even if this is difficult. In fact, you may have to fake some enthusiasm and joy around math. I do this with my own children. I genuinely love math, but I do not love a lot of the math homework they bring home, especially when it is pages of repetitive, procedural questions. But whenever my daughters say they have math homework, I always say, "Great! I love that. Let's look at it together," or something similar if they want some help or an extra eye. They now often save their math homework for me because they believe I really love it. If you don't know the math yourself, don't worry. Ask your children to explain it to you. This is a great experience for children. I often tell my own young children that I don't understand to give them the opportunity to explain it to me, and they really like to do this. Think of the birth of your own children as the perfect opportunity to start all over again with mathematics, without the people who terrorized you the first time around. I know a number of people who were traumatized by math in school but when they started learning it again as adults, they found it enjoyable and accessible. Parents of young children could make math an adult project, learning with their children or perhaps one step ahead of them each year. I teach online courses, which are available through www.youcubed.org that give parents a lot of useful information in working with their children.

Mathematical conversations should be relaxed and free from pressure. Fear and pressure impede learning, and children should always feel comfortable when offering their ideas in math. Parents and teachers should never appear irritated or judgmental if children make mistakes. Math, more than any other subject, can cause panic, which stops the mind from working.[16] I usually start any work with children by telling them that I love errors because they are really good for learning. I mean this because it is through making errors that children learn the most, as mistakes give them a chance to consider, revise, and learn new things. When I am working with children and they say something that is incorrect, I consider their thinking with them and see this as an important opportunity for learning. When students know that I am not judging them harshly and that I genuinely value errors, they are able to think more productively and learn more.

Number Flexibility

An important mission for all parents and teachers is to steer children away from the mathematical ladder of rules that I discussed in the previous chapter. Gray and Tall's research study[17] showed that successful students were those who used numbers flexibly, decomposing and recomposing them. This isn't difficult to do, but it involves children knowing that that is what they *should* be doing. Fortunately, there are specific and enjoyable ways of encouraging this number flexibility that can be used with children of all ages. One of the very best methods I know for encouraging number flexibility is that of *number talks,* the activities I introduced in the previous chapter. The aim of number talks is to get children to think of all the different ways that numbers can be calculated, decomposing and recomposing as they work. For example, you could ask a child to work out 17×5 in their heads without the use of pen and paper. This is a problem that looks difficult but becomes much easier when the numbers are flexibly moved around. So, for example, with 17×5 I could work out 15×5. I can do this in my head more easily as 10×5 is 50 and 5×5 is 25, giving me 75. I then need to remember to add 10 as I only worked out 15×5 and I need two more 5s. This gives me my answer of 85. Another way of solving the problem is not to work out 17×5 but to work out 17×10, which is 170, and then halve it. Half of 100 is 50 and half of 70 is 35, so I would get 85. As people work on problems like these, they develop number sense, which is the basis for all higher mathematics. They also develop their mental math abilities. The problem I would set children when working on number talks is to find as many ways as possible of working out the answers. Most children will find this challenging and fun.

One of the great advantages of number talk problems is that they can be posed at all sorts of levels of difficulty, and there is an endless selection of possible problems, so they can be fun for children and adults of all ages. Here are some starters of different levels of difficulty:

Addition/Subtraction	Multiplication
$25 + 35$	21×3
$17 + 55$	14×5
$23 - 15$	13×5
$48 - 17$	14×15
$56 - 19$	17×15

Some good prompts to use while you are working with children are:

- How did you think about the problem?
- What was the first step?
- What did you do next?
- Why did you do it that way?
- Can you think of a different way to do the problem?
- How do the two ways relate?
- What could you change about the problem to make it easier or simpler?

Number talks are an excellent way to teach children, of any age, to decompose and recompose numbers, which is extremely valuable in their mathematical development. But there are other great problems that require them to think creatively and use numbers flexibly. Here is a small selection:

The Four 4s

Try to make every number between 0 and 20 using only four 4s and any mathematical operation (such as multiplication, division, addition, subtraction, raising to a power, or finding a square root), with all four 4s being used each time. For example

$$5 = \sqrt{4} + \sqrt{4} + \frac{4}{4}$$

How many of the numbers between 0 and 20 can be found?

Race to 20

This is a game for two people.

Rules:
1. Start at 0.
2. Player 1 adds either 1 or 2 to 0.
3. Player 2 adds either 1 or 2 to the previous number.
4. Players continue taking turns adding 1 or 2.
5. Person who gets to 20 is the winner.

See if you can come up with a winning strategy.

Painted Cubes

A $3 \times 3 \times 3$ cube

is painted red on the outside. If it is broken up into $1 \times 1 \times 1$–unit cubes, how many of these small cubes have three sides painted? Two sides painted? One side painted? No sides painted? What about if you start with a larger original cube?

Beans and Bowls

How many ways are there to arrange 10 beans among 3 bowls? Try it for a different number of beans.

Partitions

You could use Cuisinaire rods to help with this problem.

The number 3 can be broken up into whole numbers in four different ways:

$1 + 1 + 1$

$1 + 2$

$2 + 1$

3

Or maybe you think that $1 + 2$ and $2 + 1$ are the same, so there are really only three ways to break up the number.

Decide which you like better and investigate partitions for different numbers using your rules.

In this chapter I have outlined some ways of encouraging mathematical thinking in the home or classroom through mathematical settings, puzzles, and strategies. In appendix C, I give many different resources and Web sites that can help you in your work with your own children. In the next chapter, I will summarize some critical advice for teachers, parents, and anyone who works with math learners and wants to orient them toward the most productive math learning path possible.

9 / Moving to a More Positive Future

These are exciting times, characterized by new evidence of the vast potential of all children to learn to high levels, widespread access to knowledge, incredible new technologies, and a growing movement of positivity around learning. There is increasing public awareness that the traditional mathematics education system that has been maintained for decades is badly broken, and people from across the spectrum are devoting their time and imagination to bringing about change. In the last few years, I have been fortunate to communicate this to leaders of countries as well as to teachers, district leaders, parents, CEOs, and the general public through mainstream TV shows. I have also been working with a filmmaker who has devoted her new film to the need for mathematics change. You can see the exciting and powerful film at www.youcubed.org. In my experience, I have found that every person to whom I have shown the research evidence and demonstrated the nature of true mathematics engagement (usually through video) has recognized the need for change and, in many cases, joined forces to bring it about.

Researchers of mathematics education have stood collectively and firmly for many decades to promote change. There is very little variation in the tens of thousands of studies conducted, with almost all pointing to the need for active mathematics engagement.[1] In addition to researchers, the people who lead professional development and school districts and who have had access to research knowledge have been trying to bring about the changes that I set out in this book despite the efforts of the antichange movement. But the last seven years have shown me that those who oppose change are a small, politically motivated minority, many of whom oppose an ideal that is foundational to the United States—that of equality.

We are in a time when equality is a national priority; a time with widespread awareness of the damage done by mindless math drills; and a time when those who are experts in education are being recognized and given a platform. When Cathy Williams and I formed YouCubed, we described the math change that we are promoting as a revolution. I believe that we are in the midst of an incredible revolution, one that is led by research showing that all students can reach high levels in mathematics and the nature of the teaching and parenting that brings it about.

I recently published some advice on YouCubed that I am going to share here because I believe it to be important for both teachers and parents. Teachers and parents have the opportunity to shape their children's mathematical futures. It may not seem like this is the case at times, especially when children are going through bad experiences at school or teachers are being pressured by faulty policies such as pacing guides and prescriptive curriculum. But I know, both from my extensive work with teachers and from my experience as a mother of two children, that you have the opportunity to make a huge difference in children's mathematical lives.

One of the most important contributions you can make is to dispel the idea that only some children can be successful at math or that math is a "gift" that some children have and some do not. This idea permeates American society (as well as others), but it has been completely disproved by the science of the brain and learning. The idea that some children can do well in math and some can't is a damaging myth that is harmful to children's mathematical development. All students can achieve at the highest levels of math in school if given the right opportunities and support.

Suggested Strategies

1. Never praise children by telling them they are smart. This may seem like it is encouraging but it is a fixed-ability message that is damaging.[2] When children are told they are smart they often feel good, but later when they fail in some situation—and everyone does—they think, "Hmm, I am not so smart." Instead of praising the child, always praise *what* the child has done: for example, "It is wonderful that you have learned how to add numbers," not "Wow you can add numbers, you are so smart."

When children know that learning and hard work make them achieve at the highest levels, their achievement increases significantly.[3] This idea can be hard to get across to children because mainstream media such as television programs often communicate the opposite message—that some children are smart and some are not. They communicate lots of other damaging ideas too—that math is hard, inaccessible and only for nerds. It is critical to reject these ideas as often and as loudly as you can. Instead, keep telling children that math is very exciting, and it is important to work hard because it is hard work that leads to high achievement.

2. Never share stories of math failure or even dislike. As I shared in chapter 8, research has shown that as soon as a mother says to her daughter, "I wasn't good at math at school," her daughter's achievement goes down.[4]

Even if you have to put on your best acting skills, always seem happy—even thrilled—when you see math! Don't worry if you cannot do your children's homework. Ask them to explain it to you. This can be one of the most encouraging experiences a parent can give their children. I often tell my own children that I don't know how to do the work they are doing, even if I do, because when they explain it to me, they are learning it at a much deeper level.[5]

3. Always praise mistakes and say that you are really pleased that your child is making them. Recent research has shown that our brains grow when we make mistakes.[6] Scientists have found that when people make a mistake in math, synapses spark, and there is activity in the brain that is absent when people get work correct. What this means is that we want people to make mistakes! In fact, making mistakes in math is one of the most helpful things to do. But many children (and adults) hate to make mistakes, they think it means they are not a "math person." It is important both to celebrate mistakes and to tell children their brain is growing when they make them.[7]

When my eldest daughter was young she had some terrible school experiences when teachers decided she was not smart and stopped giving her the same questions they gave to other children. As a result she developed a fixed mind-set about mathematics and was very anxious about math for some years. Now, after some careful work, she is a great math advocate and has a growth mind-set about math.

When she was ten I remember her working on two math problems, one of which she got right and one wrong. When she got the one wrong she immediately reacted badly, putting her head into her hands saying, "I can't do math." I said to her, "Do you know what just happened? When you got the one question right, nothing happened in your brain, but when you got the other question wrong, your brain grew." I give this message to my children every time they are confused, are struggling, or make a mistake. Now they tell others in their school how valuable mistakes are and how they make your brain grow.

4. Encourage children to work on problems that are challenging. We know that it is really important for students to take risks, engage in "productive struggle," and make mistakes. Sometimes my daughter asks for help with her homework when it looks difficult. I try to encourage her to have a go first without my help, saying, "I don't want to take away the opportunity for you to struggle and for your brain to grow!" Keep telling your children or students that struggle is really important. This is a delicate balance as you don't want to leave students struggling to the point that they feel despondent, but always try to encourage as much struggle as you think they can cope with at that time.

Girls, in particular, have often learned to avoid difficult work because they have more fixed mind-sets than boys—usually because they have been praised for being smart, which makes them worried about losing that label.[8] Avoidance of harder work is damaging for children and it is one of the reasons that fewer girls pursue math and science.[9]

In one of Carol Dweck's studies, participants were given

math problems that they all solved correctly. Half of the participants were praised for being "smart" and half for "working hard." When offered a choice of a follow-up problem that was easy or hard, 90 percent of the participants praised for being smart chose the easy problem, whereas most of those who were praised for working hard chose the harder problem.[10]

This tells us that the praise we give children has an immediate effect on them. It also gives us some important clues into gender inequities in mathematics participation rates.

5. When you help students, do not lead them through work step-by-step. Doing this takes away important learning opportunities for them. We often help children by solving the hard part of a problem, such as working out what the problem is asking, and then get children to do something easier, such as a calculation. For example, consider this scenario:

Question

Carlos started with 12 candies. He gave some to Janice, and then he had 8 candies. How many did he give to Janice?

> **Child:** I don't know how to do this.
> **Parent/Teacher:** Well, Carlos started with 12, and now he has 8, so what is 12 take away 8?
> **Child:** 4.
> **Parent/Teacher:** That's right!

In this scenario, the student may feel good, but the parent or teacher has done the hardest part of the problem, which is to make sense of the situation. It would be more helpful to ask the child to draw the problem out, including, if they wish, pictures of the candies. Or ask the child to restate the problem in her own words or actually manipulate objects that represent the candies. Try not to lower the cognitive demand of a problem when helping. Try not to do the hard thinking for the child, leaving her with a calculation. This, ultimately, is unhelpful for her mathematics learning.

6. Encourage drawing whenever you can. The whole of mathematics could be taught visually, which would help millions of children, but few classrooms encourage drawing, and some students believe it to be babyish. Yet mathematicians draw all the time. They do this because sketching a problem helps them really *see* the important mathematical ideas. Both drawing and restating problems help children understand what questions are asking and how the mathematics fits within them.

7. Encourage students to make sense of mathematics at all times. *Connecting Mathematical Ideas* is a book I have coauthored with Cathy Humphreys, an expert teacher who focuses on sense making at all times. In the book, we share six video cases that show Cathy's teaching as well as her lesson plans and our shared reflections on the lessons.[11]

Children should never think that math is a set of rules that they need to follow—although they often have good reason to think this! As they work on math keep asking them: "Does that make sense to you?" Why?" Or "Why not?" Discourage guessing. If children seem to be guessing, say, "Is that a guess? Because this is something we can make sense of and do not need to guess about." Mathematics is a conceptual subject, and students should be thinking conceptually at all times.[12, 13]

Here are some questions you can ask children to help them think conceptually:

- What is the question asking you?
- How could you draw this situation?
- How did you get that answer? (Ask this whether the answer is right or wrong.)
- Can you share your method with me?
- Can you try a different way of solving this?
- What does addition/probability/ratio/etc. mean?
- In what other situation could we use this?
- Would this method work with different numbers?
- What is important about this work?

8. Encourage students to think flexibly about numbers. Research has shown that the biggest difference between elementary students who are successful and those who are not is not that the higher achievers know more but that they think flexibly with numbers.[14] It is critical that children develop number sense, which means that they think flexibly with numbers and can change and regroup them. For example, a student with number sense faced with a problem like 41 minus 17 would not use an algorithm such as:

Nor would they count up from 17 or down from 41 They would change the numbers to, for example, 40 minus 16, which is a much easier question.

Often when students struggle with math early on they are given more practice with methods, facts, or skills. This is not what they need. They need a more conceptual understanding of math, and they need to develop number sense.

Many students in the United States fail algebra. The reason for this is not that algebra is really difficult but that the students lack number sense, which is the most important foundational base students can have for all higher mathematics.[15]

9. Never time children or encourage faster work. Don't use flash cards or timed tests.[16, 17, 18] Scientists can now examine brain scans when people work on math, and these have shown that timed conditions create math anxiety. Math facts are held in the working memory part of the brain, and scientists have found that when people are stressed—adults or children—their working memory becomes blocked and math facts cannot be accessed. The emphasis on speed in US math classrooms is one of the reasons we have widespread underachievement and a nation of math-traumatized people.

10. When children answer questions and get them wrong, try and find the logic in their answers—as they have usually used some logical thinking. For example if your child multiplies 3 by 4 and gets 7, don't say, "That's wrong." Instead say, "Oh, I see what you are thinking. You are using what you know about addition to add 3 and 4. When we multiply we have 4 groups of 3 . . ."

11. Give children mathematics puzzles. These have been shown to inspire children mathematically and are great for their mathematical development. Award-winning mathematician Sarah Flannery reported that her math ability and enthusiasm came not from school but from puzzles she was given to solve.[19]

12. Play games, which are similarly helpful for children's mathematical development. For young children any game with dice will help. Some board games I particularly like are:

- Place Value Safari
- Blokus
- Guess Who? (great for logical thinking)
- Mancala
- Yahtzee
- Mastermind

More games and puzzles are available on www.youcubed.org.

In addition to these twelve pieces of advice, which summarize many ideas in the book, I want to add a special message for teachers. Teachers hold students' mathematical futures in their hands, and you should never underestimate the power of your words and actions to inspire or defeat students. Almost any successful person will tell you about a teacher, usually one teacher, who believed in them and changed everything for them. You can be that teacher for all of your students.

I have talked with and been sent evidence from thousands of teachers since I started sharing the evidence and messages in this book, and I know that you can implement the changes I have described and that they can transform everything for your students. Teachers can create beautiful mathematics environments for students regardless of their past experiences or the negative district or school regulations that are imposed on them. I know teachers who are given dry repetitive math curricula to use, but they make the curricula come alive with the changes they make. Any mathematics problem can be brought to life when it is made open, more visual, and creative and is infused with students' own thinking. You can make these adaptations to problems, cut out unnecessary repetition, and invite students to go deeper into mathematical journeys with you. I am the greatest supporter of teachers, and I am fully aware from my own years as a teacher how demanding and time-consuming good teaching is. I want teachers to have rich tasks—and we are putting as many as we can on YouCubed—but I also believe in teachers' abilities to take any task and bring it to life for their students and to combine this with positive messages about students potential.

In addition to posting tasks and resources on YouCubed, I regularly post short, readable papers that summarize the research evidence we have on the ways mathematics should be taught, which you can give to administrators and parents who do not understand the need for change or who want to learn more. You are an expert on children's mathematical development and learning, and you need to use your expertise and be prepared to be strong at times to help others know the best mathematical pathways for our children to follow.

In Conclusion

Mathematics is an engaging, creative, and accessible subject that enables people to be their most powerful selves, interacting with their environment in thoughtful and informed ways. Whether you are a parent, teacher, or other education leader, you can change children's mathematics lives for the better with the knowledge we have of the brain and of powerful teaching and learning environments. We are stronger together in bringing about the changes that we need, and we should not waiver in that mission because all children deserve the very best mathematical future possible. Mathematics will play a central role in all children's lives, and together we can enable them to apply powerful, quantitative, and creative thinking to the problems they meet in their work and lives. Together we can inspire children who in turn will go on to create a brighter future for their children, filled with the scientific, creative, and technological discovery that mathematics enables. Let us move together from the mathematics trauma and dislike that has pervaded our society in recent years to a brighter mathematical future for all, charged with excitement, engagement and learning.

Viva la Revolution.

Appendix A

Solutions to the Mathematics Problems

Introduction

The Skateboard Problem

A skateboarder holds on to the merry-go-round pictured below. The platform of the merry-go-round has a 7-foot radius and makes a complete turn every 6 seconds. The skateboarder lets go at the 2 o'clock position in the picture, at which time she is 30 feet from the padded wall. How long will it take the skateboarder to hit the wall?

One Solution

The first step is to find out how far the skateboarder travels after letting go. In other words, we need to figure out the distance AB in the drawing below.

To do this we first need to work out what the angle θ is so we can use the properties of right triangles. To do this, draw the line h that passes through the center of the merry-go-round and meets the padded wall at a right angle. Since the skateboarder is at the "2-o'clock" position, which is 1/6 of the way around a clock, the angle α is 1/6 of 360°. So α = 60°. The angle β = 90°, since a tangent line of a circle always meets the radius at a right angle. Finally, α + β + θ = 180, since they are same-side interior angles between the parallel lines h and l. Therefore θ = 30°. This means that triangle ABC is a 30-60-90 triangle! Using the side relations of 30-60-90 triangles, we find:

$$BC = \frac{30}{\sqrt{3}} = 10\sqrt{3} \text{ feet}$$

and

$$AB = 20\sqrt{3} \text{ feet}$$

So now we know how far the skateboarder travels. The next step is to figure out how fast she is traveling. The merry-go-round makes a complete turn every 6 seconds. In a complete turn, the skateboarder travels the entirety of the circumference, which is

$$C = 2\pi\,(7) \approx 43.98 \text{ ft.}$$

So the skateboarder is traveling at $43.98/6 \approx 7.330$ feet per second.

Since

$$(\text{distance}) = (\text{rate})\,(\text{time})$$

then

$$(\text{time}) = (\text{distance})/(\text{rate})$$

So the time it takes the skateboarder to reach the wall is

$$\frac{20\sqrt{3}}{7.330} \approx 4.726 \text{ seconds}$$

The Chessboard Problem

Solution

What makes this problem difficult is all the different sizes of squares on a chessboard, from the smallest 1×1 squares, to overlapping 2×2 squares, all the way up to the entire chessboard, which is itself an 8×8 square.

In situations like this, it is often helpful to be organized. One way to organize the problem is to count all the different sizes of squares separately. So, let's start with the 1×1 squares. There are 8 rows and 8 columns on the board, so there are 64 of these. Next, let's look for the 2×2 squares. These are more difficult, as they can overlap, as the two gray squares below do:

Even overlapping squares have distinct center points, though, so an easy way to keep track of the overlapping squares is to mark the center of each square with a dot. Here are some examples of squares with their center points marked:

Marking all the center points of the 2 × 2 squares, we get a grid of center points:

Notice this is a 7 × 7 grid of points. So there are 49 2 × 2 squares.

To keep track of the 3 × 3 squares, we can also mark the center points, as in the following few examples:

When we draw all of the center points for the 3 × 3 squares, we get a picture that looks like this:

This is a 6 × 6 grid of points, so there are 36 3 × 3 squares. Continue this process until you have counted the 4 × 4 squares, the 5 × 5 squares, and so on. The total number of squares is then

$$8^2 + 7^2 + 6^2 + 5^2 + 4^2 + 3^2 + 2^2 + 1 = 204$$

This process of counting works for any size of chessboard. In general, for an *n*-by-*n* chessboard, the number of 1 × 1 squares is n^2. The number of 2 × 2 squares is $(n - 1)^2$ and so on. So the total number of squares is

$$n^2 + (n - 1)^2 + (n - 2)^2 + \ldots + 3^2 + 2^2 + 1$$

Chapter 3

The Railside Pattern Problem

Juan's problem: "See if you can work out how the pattern is growing and the algebraic expression that represents it!"

One Solution

A good way to solve this problem is to look at how each part is growing separately. One way to see it is this: in the diagrams on the previous page, it looks like the white squares on the left are growing by one each time the pattern proceeds to the next step, as are the white squares on the right. The solid black square on the right and the solid black square on the bottom don't seem to change. And the rectangle of gray squares is growing by one in each of its dimensions. To come up with a formula, we need to number each picture. Let's call the first picture "$n = 1$" and the next picture "$n = 2$." So now let's try to say how many of each square there are in terms of n. The white squares on the left and the white squares on the right always have one more than n, so these can each be represented by the expression $(n + 1)$. The black squares on the bottom and on the right are always just 1, no matter what n is, so these can be represented by the expression 1. Finally, the rectangle of gray squares has a width of n and a height that is two more than n, or $(n + 2)$. So the number of squares in this rectangle can be represented by the width times the height, or the expression $n(n + 2)$. So the total number of squares on the nth picture is

$$(n + 1) + (n + 1) + 1 + 1 + n(n + 2)$$
$$= n + 1 + n + 1 + 1 + 1 + n^2 + 2n$$
$$= n^2 + 4n + 4$$

Interesting note: this expression factors as the perfect square, $(n + 2)^2$, which means that every arrangement of squares can be arranged as a square. Try to see how they rearrange. This could lead to a different way of solving the problem.

The Amber Hill Question

Helen rides a bike for 1 hour at 30 km/hour and 2 hours at 15 km/hour. What is Helen's average speed for the journey?

Solution

"Average speed" is one of those tricky expressions in word problems, because it can be interpreted to mean different things. The most natural interpretation is "If you were traveling at a *constant* speed, how fast would you be going to cover the same distance in the same amount of time?" This interpretation allows you to work out the total distance traveled and the total time spent. Then the average speed is (total distance) / (total time). In this problem, Helen travels 30 km for the first hour and $15 \times 2 = 30$ km for the second and third hours. So the total distance is 30 km + 30 km = 60 km. The total time is 1 hour + 2 hours = 3 hours. So the average speed is (total distance) / (total time) = (60 km) / (3 hours) = 20 km/hour.

Chapter 7

The Staircase Problem

In this task students were asked to determine the total number of blocks in a staircase that grew incrementally from 1 block high, to 2 blocks high, to 3 blocks high, and so on, as a move toward predicting a 10-block–high staircase and a 100-block–high staircase, and, finally, algebraically expressing the total blocks in any staircase. Students were provided with a box full of linking cubes to build the staircases if they wished.

A four-block-tall staircase: total blocks = 4 + 3 + 2 + 1 = 10

Solution

There are many ways to "see" the growth in such a staircase. One of the most elegant is to think about two copies of the staircase fitted together, as shown here:

These fit together perfectly into a 4 × 5 rectangle. Since we put two staircases together, the total number of squares in our original staircase is (4 × 5)/2 = 10. In general, two of the *n*th staircases can be put together to make an $n(n + 1)$ rectangle. Again, since two staircases were put together, the total number of squares in one is $n(n + 1)/2$. This is sometimes called the *n*th triangular number, because a staircase is triangular in shape. Using this formula we can work out how many squares are in each staircase. For example, the 10th staircase ($n = 10$) has 10(10 + 1)/2 = 55 squares. The 100th staircase has 100(100 + 1)/2 = 5050 squares. Rumor has it the famous mathematician Carl Gauss worked out in this way when his teacher asked the class to add up the numbers 1 through 100. Can you work out why this would give you the correct answer for that sum?

Alonzo's Staircase Problem

5 blocks   *5 + 9 = 14 blocks*   *5 + 9 + 13 = 27 blocks*

Solution

Alonzo's problem is like the staircase problem, except the staircase comes out in four directions from a center point. Again, we have numerous ways to count how many squares there are in this shape. One method is to use the solution to the staircase problem. Each of Alonzo's staircases is formed by 4 copies of the staircases from the above problem, plus the squares in the center column, where the 4 copies are attached. So the number of squares is $4n[(n + 1)/2] + n$. The first term represents the 4 staircases, and the n is the column of squares in the middle. This last term is n because Alonzo's shape is 1 high in the first case, 2 high in the second case, and in general n high in the nth case. This formula simplifies as follows:

$$4n[(n + 1)/2] + n$$
$$= 2n(n + 1) + n$$
$$= 2n^2 + 2n + n$$
$$= 2n^2 + 3n$$

Whenever you come up with a general algebraic formula, it's good to make sure it works for the small cases that you understand. Can you try this formula to see if it works for the initial cases of Alonzo's staircase pictured? Can you come up with this formula by grouping the cubes a different way?

Cowpens & Bullpens Problem

During an activity called "Cowpens & Bullpens" students had to determine how many lengths of fencing were required to contain an increasing number of cows, given certain fencing parameters.

Solution

There are many ways to see this problem's pattern, as with others in the book. One way is to count the number of sides of fencing on each side of the cows, then add the corners. The first interesting thing to notice is that the number of sides of fencing above and below the cows is the same as the number of cows, and the number of sides of fencing to the left and right is always 1. So for the first picture above there are $4 + 4 + 1 + 1 + 4 = 14$ sides of fencing, where the first two 4s come from the fencing above and below the cows, the 1s come from the fencing to the left and the right of the cows, and the last 4 comes from the 4 corners. For the next picture there are $5 + 5 + 1 + 1 + 4$ for similar reasons. In general, there are $n + n + 1 + 1 + 4$ sides of fencing, or $2n + 6$ sides.

Chapter 8
Problems from Sarah Flannery's Book

The Two-Jars Puzzle

Given a 5-liter jar and a 3-liter jar and an unlimited supply of water, how do you measure out 4 liters exactly?

Solution

There are many ways to solve this problem—actually an infinite number! Here's one way. Fill the 5-liter jar. Pour the water from the 5-liter jar into the 3-liter jar until the 3-liter jar is full. Now you have 2 liters of water remaining in the 5-liter jar. Dump out the water in the 3-liter jar. Then, put the 2 liters of water from the 5-liter jar into the 3-liter jar. Now fill up the 5-liter jar completely. Pour water from the 5-liter jar into the 3-liter jar until it is full. Since there were already 2 liters of water in the 3-liter jar, you have poured exactly 1 liter out of the 5-liter jar. So, there are exactly 4 liters remaining in the 5-liter jar.

If that was confusing, here's a table showing how many liters of water are in each jar at each step, with explanations of what happened at each step:

3-liter jar	5-liter jar	What happened
empty	empty	Nothing yet!
empty	5 liters	Filled the 5-liter jar to the brim.
3 liters	2 liters	Filled the 3-liter jar with the 5-liter jar.
empty	2 liters	Dumped out the 3-liter jar.
2 liters	empty	Poured the contents of the 5-liter jar into the 3-liter jar.
2 liters	5 liters	Filled the 5-liter jar to the brim.
3 liters	**4 liters!**	Topped off the 3-liter jar using contents of the 5-liter jar.

An interesting follow-up question: Is this the least number of steps, or is there a faster way to measure 4 liters? Another follow-up question: Are there any quantities you can't make with these two jars?

The Rabbit Puzzle

A rabbit falls into a dry well 30 meters deep. Since being at the bottom of a well was not her original plan, she decides to climb out. When she attempts to do so, she finds that after going up 3 meters (and this is the sad part), she slips back 2. Frustrated, she stops where she is for that day and resumes her efforts the following morning—with the same result. How many days does it take her to get out of the well?

Solution

This is a classic example of a "trick" problem. Even if you see the trick, it is easy to make a mistake. One good thing to notice is that the act of climbing and sliding can be greatly simplified. Instead of thinking of it as going up 3 meters and down 2 meters every day, you can just think about it as going up 1 meter. So every day the rabbit goes up 1 meter— which should mean it takes her 30 days to get out, 1 meter each day. But the reason it does not is that on the last day the rabbit actually gets out, she doesn't slip back down 2 meters. So the rabbit saves itself 2 days, and it only takes 28 days. Can you work out other ways of saying this and seeing why it is actually 28 days?

The Buddhist Monk Puzzle

One morning, exactly at sunrise, a Buddhist monk leaves his temple and begins to climb a tall mountain. The narrow path, no more than a foot or two wide, spiraled around the mountain to a glittering temple at the summit. The monk ascended the path at varying rates of speed, stopping many times along the way to rest and eat the dried fruit he carried with him. He reached the temple shortly before sunset. After several days of fasting he begins his journey back along the same path, starting at sunrise and again walking at variable speeds with many pauses along the way, finally arriving at the lower temple just before sunset. Prove that there is a spot along the path that the monk will occupy on both trips at precisely the same time of day.

Solution

This problem is part of a beautiful class of problems called "fixed-point theorems." If you like this one, there are many others like it! One of the prettiest ways to see this solution is to imagine the graph of the monk's journey, with time on the x-axis, and position on the y-axis. So his journey on the first day might look something like this:

Then, on the same graph we can represent his journey down the mountain:

Note, these two paths may look very different, because he might choose to speed up or slow down at different times. However, since the first path must go from the lower left corner to the upper right corner, and the second path must go from the upper left corner to the lower right corner, they must cross somewhere. And, as you can see, they do cross somewhere. This point marks the time of day and the location in which the monk was in the same place at the same time on both days.

The Four 4s

Try to make every number between 0 and 20 using only four 4s and any mathematical operation (such as multiplication, division, addition, subtraction, raising to a power, or finding a square root), with all four 4s being used each time. For example

$$5 = \sqrt{4} + \sqrt{4} + \frac{4}{4}$$

How many of the numbers between 0 and 20 can be found?

Solution

There are many ways to make some numbers with four 4s, but for some other numbers it is much more difficult. Here's one set of solutions for the numbers 0 through 20:

$$0 = 4 - 4 + 4 - 4$$
$$1 = 4/4 + 4 - 4$$
$$2 = 4/4 + 4/4$$
$$3 = \sqrt{4 \times 4} - 4/4$$
$$4 = \sqrt{4} + \sqrt{4} + 4 - 4$$
$$5 = \sqrt{4} + \sqrt{4} + 4/4$$
$$6 = 4 + \sqrt{4} + 4 - 4$$
$$7 = 4 + \sqrt{4} + 4/4$$
$$8 = 4\sqrt{4} + 4 - 4$$
$$9 = 4 + 4 + 4/4$$
$$10 = 4 \times 4 - 4 - \sqrt{4}$$
$$11 = \frac{\sqrt{4}(4! - \sqrt{4})}{4}$$
$$12 = 4(4 - 4/4)$$

$$13 = \frac{\sqrt{4}(4! + \sqrt{4})}{4}$$
$$14 = 4! - 4 - 4 - \sqrt{4}$$
$$15 = 4 \times 4 - 4/4$$
$$16 = 4 \times 4 + 4 - 4$$
$$17 = 4 \times 4 + 4/4$$
$$18 = 4! - \sqrt{4} - \sqrt{4} - \sqrt{4}$$
$$19 = 4! - 4 - 4/4$$
$$20 = 4 \times (4 + 4/4)$$

Notice for some solutions, I used the factorial notation, the exclamation point (!), after some numbers. This operation multiplies the number by every positive integer less than it. So $4! = 4 \times 3 \times 2 \times 1 = 24$. For the numbers where I used a factorial operation, do you think it's possible to find solutions without it?

Race to 20

This is a game for two people.

Rules:
1. Start at 0.
2. Player 1 adds either 1 or 2 to 0.
3. Player 2 adds either 1 or 2 to the previous number.
4. Players continue taking turns adding 1 or 2.
5. The person who gets to 20 is the winner.

See if you can come up with a winning strategy.

Solution
One of the things you may notice after playing this game for a while is that if you can get to 17, you are the winner. This is because no matter what your opponent adds, whether it be 1 or 2, on your next turn you will be able to get to 20. So getting to 17 is just as good as getting to

20. This idea can extend to even lower numbers. Here, 17 is a good number for you to get to because it is 3 away from 20, just one more than your opponent can add. Keeping 3 away from these "good numbers" is the trick. By similar reasoning, getting 14 makes you the winner, because no matter what your opponent adds, on your next turn you will be able to get to 17, and then you already know what you need to do to get to 20. Similar reasoning applies to 11, 8, 5, and so on, down by 3s. So now start at the beginning: Can you come up with a winning strategy if you go first? What about if you go second? If you like, make up some variations of this game that work differently, then try to find the strategy.

Painted Cubes

A $3 \times 3 \times 3$ cube

is painted red on the outside. If it is broken up into $1 \times 1 \times 1$–unit cubes, how many of these small cubes have three sides painted? Two sides painted? One side painted? No sides painted? What about if you start with a larger original cube?

Solution

The 3×3 cube has 1 cube right in the middle, with 0 sides painted. It has 6 cubes with 1 side painted, 1 in the center of each of the six faces. It has 12 cubes with 2 sides painted, 1 in the center of each edge of the large cube. And it has 8 cubes with 3 sides painted, 1 in each of the 8 corners of the large cube.

In general, for an $n \times n \times n$ cube, let's think about how to count all the cubes with 0 sides painted. Imagine removing the entire outer layer of small cubes. You'll be left with a cube in the center, but each of its dimensions will be shrunk by 2, because a layer of cubes has been removed on both sides. So it's an $n - 2$ by $n - 2$ by $n - 2$ cube. So it has $(n - 2)^3$ little cubes. For the cubes with 1 side painted, these are on the interior of each face. By similar reasoning as above, this is an $n - 2$ by $n - 2$ square, so there are $(n - 2)^2$ of these for each of the 6 faces, so $6(n - 2)^2$ have 1 side painted. For the cubes with 2 sides painted, these are along each of the 12 edges, but there are $n - 2$ of these (can you see why?), making for a total of $12(n - 2)$ of these. Finally, no matter how large the cube is, there is only one cube with 3 sides painted per each of the 8 corners, so there are 8 cubes with three sides painted.

Beans and Bowls

How many ways are there to arrange 10 beans among 3 bowls? Try it for different numbers of beans.

Solution

There are lots of ways to do this problem. One way is to break it into 11 cases, based on how many beans are in the first bowl. (Can you see why it is 11 cases and not 10 cases?) Then for each case, count how many ways the beans can be distributed among the other 2 bowls. I recommend doing this, so that you understand the problem more by immersing yourself in it. Once you have done that, here's a quicker and more elegant way to see the final answer:

Imagine your beans as dots, all lined up, but with 2 extras:

● ● ● ● ● ● ● ● ● ● ● ●

Why two extras? Well, imagine that it's your job to replace two of the beans with an x, like this:

$$\bullet \bullet x \bullet \bullet \bullet x \bullet \bullet \bullet \bullet \bullet$$

or this:

$$\bullet \bullet \bullet \bullet \bullet \bullet \bullet x \bullet x \bullet \bullet$$

or even this:

$$\bullet \bullet \bullet \bullet x x \bullet \bullet \bullet \bullet \bullet \bullet$$

These x's are instructions on which bowls to put the beans into. They work as follows: put beans into the first bowl until you hit the first x; then put beans into the second bowl until you hit the second x; then put the rest of the beans into the third bowl. So in the first example above, there would be 2 beans in the first bowl, 3 beans in the second bowl, and 5 beans in the third bowl. For the second example, there would be 7 beans in the first bowl, 1 bean in the second bowl, and 2 beans in the third bowl. For the last example, there would be 4 beans in the first bowl, 0 beans in the second bowl (do you see why?), and 6 beans in the third bowl.

So each way of replacing 2 of the 12 beans with an x corresponds to a way to put the remaining 10 beans in bowls. So if we can figure out how many ways there are to choose 2 beans and replace them with an x, then we will know how many arrangements of beans in bowls there are. How many ways are there to pick the first bean? Well, there are 12, because you can pick any bean. Then there are 11 ways to pick the second bean, because you've already picked 1. So there are 12 × 11 = 132 ways to pick 1 bean and then pick another bean. But, there is another thing we need to be careful of. There are 2 ways to pick each pair of x's, because you can switch the order in which you picked them. So 132 counts every pair of x's twice, and we need to divide by 2. So the number of ways of replacing 2 beans by x's, which is the same as the number of ways to put 10 beans in 3 bowls, is (12 × 11)/2 = 66.

Can you see how this method would work if you changed the number of beans? What about if you changed the number of bowls?

Partitions

You could use Cuisenaire rods to help with this problem.

The number 3 can be broken up into positive numbers in four different ways:

1 + 1 + 1

1 + 2

2 + 1

3

Or maybe you think that 1 + 2 and 2 + 1 are the same, so there are really only three ways to break up the number.

Decide which you like better and investigate partitions for different numbers using your rules.

Solution

For this one, you're on your own! Coming up with a general pattern for how many partitions a number has is an unsolved problem. Welcome to cutting-edge mathematics.

Appendix B

Recommended Curriculum

K–5

- TERC: Investigations (Grades K–5). Publisher: Pearson Scott Foresman. https://investigations.terc.edu.
- ST Math. Publisher: MIND Research. www.mindresearch.org.
- Bridges (Grades K–5). Publisher: The Math Learning Center. www.mathlearningcenter.org/bridges.
- Contexts for Learning.

6–8

- College Preparatory Mathematics (CPM) Core Connections Series. Publisher: CPM Educational Program. www.cpm.org.
- Connected Mathematics Project. Publisher: Michigan State University. https://connectedmath.msu.edu.

Intervention/Supplemental

- ST Math. Supplemental curriculum. Publisher: MIND Research. www.mindresearch.org.
- Math 180. Publisher: Scholastic. http://teacher.scholastic.com/products/math180.

9–12

- Core-Plus Mathematics, CPMP. Publisher: McGraw-Hill. http://wmich.edu/cpmp.
- CPM Core Connections Series. Publisher: CPM Educational Program. www.cpm.org.
- Interactive Mathematics Program, IMP. Publisher: It's About Time Interactive. http://mathimp.org.

Appendix C

Further Resources

Books

Boaler, Jo. *Mathematical Mindsets: Unleashing Students' Potential Through Creative Math, Inspiring Messages and Innovative Teaching.* Hoboken, NJ: Wiley, 2015.

Boaler, Jo, and Cathy Humphreys. *Connecting Mathematical Ideas: Middle School Video Cases to Support Teaching and Learning.* Portsmouth, NH: Heinemann, 2005.

Cohen, E. G., and R. A. Lotan. *Designing Groupwork: Strategies for the Heterogeneous Classroom.* New York: Teachers College Press, 2014.

Driscoll, M. *Fostering Algebraic Thinking: A Guide for Teachers, Grades 6–10.* Portsmouth, NH: Heinemann, 1999.

Driscoll, M., R. W. DiMatteo, J. Nikula, and M. Egan. *Fostering Geometry Thinking: A Guide for Teachers, Grades 5–10.* Portsmouth, NH: Heinemann, 2007.

Harris, P. W. *Building Powerful Numeracy for Middle and High School Students.* Portsmouth, NH: Heinemann, 2011.

Hiebert, J., T. P. Carpenter, E. Fennema et al. *Making Sense: Teaching and Learning Mathematics with Understanding.* Portsmouth, NH: Heinemann, 1997.

Humphreys, Cathy, and R. Parker. *Making Number Talks Matter: Developing Mathematical Practices and Deepening Understanding, Grades 4–10.* Portland, ME: Stenhouse, 2015.

Lamon, S. J. *Teaching Fractions and Ratios for Understanding: Essential Content Knowledge and Instructional Strategies for Teachers.* New York: Routledge, 2011.

Mason, J. *Learning and Doing Mathematics*. Hertfordshire, England: Tarquin, 2000.

Mason, J., L. Burton, and K. Stacey. (2010). *Thinking Mathematically*. Upper Saddle River, N.J.: Pearson, 2010.

Moses, R. P. and C. E. Cobb Jr. *Radical Equations: Math Literacy and Civil Rights*. Boston: Beacon Press 2001.

Parrish, S. *Number Talks: Helping Children Build Mental Math and Computation Strategies, Grades K–5, Updated with Common Core Connections*. Sausalito, CA: Math Solutions, 2014.

Petersen, J. *Math Games for Independent Practice: Games to Support Math Workshops and More*. Sausalito, CA: Math Solutions, 2013.

Pólya, G. *How to Solve It: A New Aspect of Mathematical Method*. Princeton, N.J.: Princeton University Press, 2014.

R4 Educated Solutions. *Making Math Accessible to English Language Learner: Practical Tips and Suggestions*. Bloomington, IN: Solution Tree, 2009.

Shumway, J. F. *Number Sense Routines: Building Numerical Literacy Every Day in Grades K–3*. Portland, ME: Stenhouse., 2011

Small, M. *Good Questions: Great Ways to Differentiate Mathematics Instruction*. New York: Teachers College Press, 2012.

Small, M., and A. Lin. More Good Questions: Great Ways to Differentiate Secondary Mathematics Instruction. New York: Teachers College Press, 2010.

Stein, M. K., M. S. Smith, M. A. Henningsen, and E. A. Silver. *Implementing Standards-Based Mathematics Instruction: A Casebook for Professional Development, Second Edition* by Mary Kay Stein et al. New York: Teachers College Press, 2009.

Swetz, F. J. *Mathematical Expeditions: Exploring World Problems Across the Ages*. Baltimore: Johns Hopkins University Press, 2012.

Tate, M. Worksheets Don't Grow Dendrites: 20 Instructional Strategies That Engage The Brain. Thousand Oaks, CA: Corwin, 2010.

Wiggins, G., and J. McTighe. *Understanding by Design*. Upper resources). Saddle River, NJ: Pearson, 2005.

Wiliam, D. *Embedded Formative Assessment*. Bloomington, IN: Solution Tree, 2011.

Web Sites

Balanced Assessment: http://balancedassessment.concord.org

Conceptua Math (grades 3–6): Visual and conceptual learning, www.conceptuamath.com

Dan Meyer's resources: http://blog.mrmeyer.com

Estimation 180: www.estimation180.com

GeoGebra: http://geogebra.org/cms

Hour of Code: http://hourofcode.com/us

Mathalicious (grades 6–12): Real-world lessons for middle and high school, www.mathalicious.com

The Math Forum: www.mathforum.org.

The Mathematics Assessment Project, Shell Center: http://map.math shell.org/materials/index.php

National Council of Teachers of Mathematics (NCTM): www.nctm.org (You need to be a member to access some of the resources.)

NCTM Illuminations: http://illuminations.ntcm.org

NRICH: http://nrich.maths.org

Number Strings: http://numberstrings.com

Shell Centre for Mathematical Education Publications: www.mathshell.com

Teach the Hour of Code: http://code.org

Video Mosaic Collaborative: http://videomosaic.org

Visual Patterns (grades K–12): www.visualpatterns.org

YouCubed: www.youcubed.org

Apps

ABCya.com: Virtual Manipulatives

Bedtime Math Foundation: Bedtime Math

BrainQuake Inc.: Wuzzit Trouble.

DragonBox: Algebra 5+, Algebra 12+, Elements

Educreations

Explain Everything

MicroBlink: PhotoMath

MIND Research Institute: BigSeed and KickBox

Motion Math: Zoom, Hungry Fish, and Fractions!

Notability

The Math Learning Center: Number Rack

Scholastic: Sushi Monster

TapTapBlocks

Wolfram Group LLC: Wolfram|Alpha

Games

Calculation Nation: http://calculationnation.nctm.org
Mathbreakers: www.mathbreakers.com
Minecraft: https://minecraft.net

Puzzle Books

Berlekamp, E., and T. Rodgers. *The Mathemagician and Pied Puzzler: A Collection in Tribute to Martin Gardner.* Natick, MA: AK Peters, 1999.

Berlekamp, E. R., J. H. Conway, and R. K. Guy. *Winning Ways for Your Mathematical Plays,* 2nd ed. Wellesley, MA: AK Peters, 2001.

Bolt, B. *The Amazing Mathematical Amusement Arcade.* Cambridge, England: Cambridge University Press, 1984.

———. *Mathematical Cavalcade.* Cambridge, England: Cambridge University Press, 1992.

———. *A Mathematical Pandora's Box.* Cambridge, England: Cambridge University Press, 1993.

———. *A Mathematical Jamboree.* Cambridge, England: Cambridge University Press, 1995.

Cornelius, M., and A. Parr. *What's Your Game?* Cambridge, England: Cambridge University Press, 1991.

Gardner, M. *Riddles of the Sphinx: And Other Mathematical Puzzle Tales.* Washington, DC: Mathematical Association of America, 1987.

———. *Mathematical Puzzle Tales.* Washington, DC: Mathematical Association of America, 2000.

———. *The Colossal Book of Mathematics.* New York: W.W. Norton, 2001.

Moscovich, I. *Knotty Number Problems & Other Puzzles.* New York: Sterling, 2005.

Tanton, J. *Solve This: Math Activities for Students and Clubs.* Washington, DC: Mathematical Association of America, 2001.

Notes

Preface to the New Edition

1. Keith Devlin, *What Is Mathematics? Introduction to Mathematical Thinking* (Spring 2014), https://www.coursera.org/course/maththink.

2. Jo Boaler, "Research Suggests That Timed Tests Cause Math Anxiety," *Teaching Children Mathematics* 20, no. 8 (April 2014): 469–74. Available for download at www.youcubed.stanford.edu.

3. Jo Boaler, "Ability and Mathematics: The Mindset Revolution That Is Reshaping Education," *Forum* 55, no. 1 (2013): 143–52. Available for download at www.youcubed.stanford.edu.

4. Jo Boaler, "Fluency Without Fear: Research Evidence on the Best Ways to Learn Math Facts" (January 2014). Available for download at www.youcubed.stanford.edu.

5. "How to Learn Math: For Students," Stanford University, http://online.stanford.edu/course/how-to-learn-math-for-students-s14.

6. Odula O. Abiola and Hakirat S. Dhindsa, "Improving Classroom Practices Using Our Knowledge of How the Brain Works," *International Journal of Environmental & Science Education* 7, no. 1 (January 2012): 71–81.

7. "PISA 2012 Results," OECD, http://www.oecd.org/pisa/keyfindings/pisa-2012-results.htm.

8. Katherine Woollett and Eleanor A. Maguire, "Acquiring 'the Knowledge' of London's Layout Drives Structural Brain Changes," *Current Biology* 21,

no. 24 (December 2011): 2109–14, http://www.cell.com/current-biology/abstract/S0960-9822(11)01267-X.

9. Boaler, "Ability and Mathematics."

10. Jason S. Moser, Hans S. Schroder, Carrie Heeter, T. P. Moran, and Yu-Hao Le, "Mind Your Errors: Evidence for a Neural Mechanism Linking Growth Mind-set to Adaptive Posterror Adjustments," *Psychological Science* 22, (2011): 1484–89.

11. http://online.stanford.edu/course/how-to-learn-math-for-students-s14.

12. Elena Silva and Taylor White, *Pathways to Improvement: Using Psychological Strategies to Help College Students Master Developmental Math*. The Carnegie Foundation for the Advancement of Teaching, 2013.

13. Lawyers' Committee for Civil Rights of the San Francisco Bay Area (LCCR), *Held Back: Addressing Misplacement of 9th Grade Students in Bay Area School Math Classes*, January 2013, http://www.lccr.com/wp-content/uploads/Held_Back_9th_Grade_Math_Misplacement.pdf.

Introduction: Understanding the Urgency

1. Programme for International Student Assessment (PISA), *Learning from Tomorrow's World: First Results from PISA 2003* (Italy: Organisation for Economic Co-operation and Development [OECD], 2003).

2. "Chapter 2. Higher Education in Science and Engineering," National Science Foundation: Science and Engineering Indicators 2014, http://www.nsf.gov/statistics/seind14/index.cfm/chapter-2.

3. "Science, Technology, Engineering, and Math: Education for Global Leadership," US Department of Education, http://www.ed.gov/stem.

4. MathAlive! *Math Relevance to U.S. Middle School Students: A Survey Commissioned by Raytheon Company*, 2012, http://www.mathmovesu.com/sites/default/files/Math-Relevance_rtn12_studentsmth_results_2012.pdf.

5. Silva and White, *Pathways to Improvement*.

6. Susan L. Forman and Lynn A. Steen, *Beyond Eighth Grade: Functional Mathematics for Life and Work* (Berkeley: National Center for Research in Vocational Education, University of California, 1999), 14–15.

7. "Conrad Wolfram: Teaching Kids Real Math with Computers," TED, http://www.ted.com/talks/conrad_wolfram_teaching_kids_real_math_with_computers?language=en.

8. Robert P. Moses and Charles E. Cobb Jr., *Radical Equations: Math Literacy and Civil Rights* (Boston: Beacon Press, 2001).

9. Lynn A. Steen, ed., *Why Numbers Count: Quantitative Literacy for Tomorrow's America* (New York: College Entrance Examination Board, 1997).

10. Julie Gainsburg, *The Mathematical Behavior of Structural Engineers* (Stanford University, 2003), *Dissertation Abstracts International* A 64, no. 5: 34.

11. Celia Hoyles, Richard Noss, and Stefano Pozzi, "Proportional Reasoning in Nursing Practice," *Journal for Research in Mathematics Education* 32, no. 1 (2001): 4–27.

12. Richard Noss, Celia Hoyles, and Stefano Pozzi, "Abstraction in Expertise: A Study of Nurses' Conceptions of Concentration," *Journal for Research in Mathematics Education* 33, no. 3 (2002): 204–29.

13. Gainsburg, *Mathematical Behavior*, 36.

14. "How to Learn Math: For Teachers and Parents," Stanford Graduate School of Education, http://scpd.stanford.edu/instanford/how-to-learn-math.jsp.

15. J. Lave, M. Murtaugh, and O. de la Rocha, "The Dialectical Construction of Arithmetic Practice," in Barbara Rogoff and Jean Lave, eds., *Everyday Cognition: Its Development in Social Context* (Cambridge, MA: Harvard University Press, 1984).

16. Jean Lave, *Cognition in Practice: Mind, Mathematics and Culture in Everyday Life* (Cambridge, England: Cambridge University Press, 1988).

17. Jo Masingila, "Learning from Mathematics Practice in Out-of-School Situations," *For the Learning of Mathematics* 13, no. 2 (1993): 18–22.

18. Terezinha Nunes, Analucia D. Schliemann, and David W. Carraher, *Street Mathematics and School Mathematics* (New York: Cambridge University Press, 1993).

19. Lave, *Cognition in Practice*.

20. Jo Boaler, *Experiencing School Mathematics: Traditional and Reform Approaches to Teaching and Their Impact on Student Learning*, revised and expanded edition (Mahwah, NJ: Lawrence Erlbaum, 2002).

21. Carolyn Maher, "Is Dealing with Mathematics as a Thoughtful Subject Compatible with Maintaining Satisfactory Test Scores? A Nine-Year Study," *Journal of Mathematical Behavior* 10, no. 3 (1991): 225–48.

22. Boaler, *Experiencing School Mathematics*.

23. All the names of schools, teachers, and students in this book are pseudonyms. Schools that are involved in research studies are always promised complete anonymity, a requirement of University Institutional Review Boards.

1/What Is Math? And Why Do We *All* Need It?

1. Dan Brown, *The Da Vinci Code* (New York: Doubleday, 2003).

2. Margaret Wertheim, *Pythagoras's Trousers: God, Physics, and the Gender Wars* (New York: W. W. Norton, 1997), 3–4.

3. Boaler, *Experiencing School Mathematics*.

4. Keith Devlin, *The Math Gene: How Mathematical Thinking Evolved and Why Numbers Are Like Gossip* (New York: Basic Books, 2000), 7.

5. Patricia C. Kenschaft, *Math Power: How to Help Your Child Love Math, Even If You Don't,* revised edition (Upper Saddle River, NJ: Pi Press, 2005).
6. W. W. Sawyer, *Prelude to Mathematics* (New York: Dover, 1955), 12.
7. Nicholas Fiori, "In Search of Meaningful Mathematics: The Role of Aesthetic Choice," Stanford PhD dissertation, http://searchworks.stanford.edu/view/6969509.
8. Simon Singh, *Fermat's Enigma: The Epic Quest to Solve the World's Greatest Mathematical Problem* (New York: Anchor Books, 1997).
9. Ibid.
10. Ibid., xiii.
11. Quoted in ibid., 6.
12. Fran Schumer, "Jersey; In Princeton, Taking On Harvard's Fuss About Women," *New York Times,* June 19, 2005.
13. Imre Lakatos, *Proofs and Refutations: The Logic of Mathematical Discovery* (Cambridge, England: Cambridge University Press, 1976).
14. W. H. Cockcroft, *Mathematics Counts: Report of Inquiry into the Teaching of Mathematics in Schools* (London: Her Majesty's Stationery Office, 1982).
15. Robin Wilson, in Donald J. Albers, Gerald L. Alexanderson, and Constance Reid, *More Mathematical People: Contemporary Conversations* (Boston: Harcourt Brace Jovanovich, 1990), 30.
16. Devlin, *Math Gene,* 76.
17. George Pólya, *How to Solve It: A New Aspect of Mathematical Method* (New York: Doubleday Anchor, 1971), v.
18. Leone Burton, "The Practices of Mathematicians: What Do They Tell Us About Coming to Know Mathematics?" *Educational Studies in Mathematics* 37 (1999): 36.
19. Peter Hilton, popular quote.
20. Reuben Hersh, *What Is Mathematics, Really?* (New York: Oxford University Press, 1997), 18.
21. Devlin, *Math Gene,* 9.
22. Jo Boaler, "Britain's Maths Policy Simply Doesn't Add Up," *Telegraph,* August 14, 2014, http://www.telegraph.co.uk/education/educationnews/11031288/Britains-maths-policy-simply-doesnt-add-up.html.
23. Fiori, *Practices of Mathematicians.*
24. Boaler, *Experiencing School Mathematics.*
25. David A. Reid, *Conjectures and Refutations in Grade 5 Mathematics* (doctoral dissertation, Indiana University, 2002).
26. Gloria E. Keil, *Writing and Solving Original Problems as a Means of Improving Verbal Arithmetic Problem Solving Ability* (unpublished doctoral dissertation, Indiana University, 1965).

2/What's Going Wrong in Classrooms?
Identifying the Problems

1. Alan H. Schoenfeld, "The Math Wars," *Educational Policy* 18, no. 1 (2004): 253–86.
2. William L. Sanders and June C. Rivers, *Cumulative and Residual Effects of Teachers on Future Student Academic Achievement* (Knoxville: University of Tennessee, Value-Added Research and Assessment Center, 1996).
3. S. Paul Wright, Sandra P. Horn, and William L. Sanders, "Teacher and Classroom Context Effects on Student Achievement: Implications for Teacher Evaluation," *Journal of Personnel Evaluation in Education* 11, no. 1 (1997): 57–67.
4. Heather R. Jordan, Robert L. Mendro, and Dash Weersinghe, *Teacher Effects on Longitudinal Student Achievement: A Preliminary Report on Research on Teacher Effectiveness* (paper presented at the National Evaluation Institute, Indianapolis; Kalamazoo, MI: Western Michigan University, 1997).
5. Linda Darling-Hammond, "Teacher Quality and Student Achievement: A Review of State Policy Evidence," *Educational Policy Analysis Archives* 8, no. 1 (2000).
6. L. Rosen, "Calculating Concerns: The Politics of Representation in California's 'Math Wars,'" University of California, San Diego.
7. Richard G. Brown, Mary P. Dolciani, Robert H. Sorgenfrey, and William L. Cole, *Algebra: Structure and Method* (Evanston, IL: Houghton Mifflin, McDougal Littell, 2000), 1.
8. Ibid., 3.
9. Daniel Fendel, Diane Resek, Lynne Alper, and Sherry Fraser, *Interactive Mathematics Program* (Emeryville, CA: Key Curriculum Press, 1997), 189.
10. Ibid., 194.
11. Ibid., 203–4.
12. Wayne Bishop, "Another Terrorist Threat," *Math Forum,* March 11, 2003.
13. "Professor Jo Boaler," Stanford University, http://stanford.edu/~joboaler/.
14. Scott Jaschik "Casualty of the Math Wars," Inside Higher Ed, October 15, 2012, https://www.insidehighered.com/news/2012/10/15/stanford-professor-goes-public-attacks-over-her-math-education-research.
15. "Programme for International Student Assessment (PISA)," OECD, http://www.oecd.org/pisa/.
16. E. G. Johnson and N. L. Allen, *The NAEP 1990 Technical Report* (No. 21-TR-20), Washington, DC: National Center for Educational Statistics, 1992.
17. Jo Boaler, *Experiencing School Mathematics: Teaching Styles, Sex and Setting* (Buckingham, England: Open University, 1997).

18. Thomas P. Carpenter, Megan L. Franke, and L. Levi, *Thinking Mathematically: Integrating Arithmetic & Algebra in the Elementary School* (Portsmouth, NH: Heinemann, 2003).

19. Sarah Flannery, *In Code: A Mathematical Journey* (Chapel Hill, NC: Algonquin Books, 2002), 38.

20. Hersh, *What Is Mathematics, Really?*, 27.

21. Jo Boaler and Jim Greeno, "Identity, Agency and Knowing in Mathematics Worlds," in Jo Boaler, ed., *Multiple Perspectives on Mathematics Teaching and Learning* (Westport, CT: Ablex, 2000), 171–200.

22. A. Murata, personal communication, 2006.

23. Alan H. Schoenfeld, "Confessions of an Accidental Theorist," *For the Learning of Mathematics* 7, no. 1 (1987): 37.

24. Hilary Rose, "Reflections on PUS, PUM and the Weakening of Panglossian Cultural Tendencies," *The Production of a Public Understanding of Mathematics* (Birmingham, England: University of Birmingham, 1998), 4.

25. Ibid., 5.

3/A Vision for a Better Future:
Effective Classroom Approaches

1. "Program for Complex Instruction," Stanford University, http://cgi.stan ford.edu/group/pci/cgi-bin/site.cgi.

2. Leslie S. Siskin, *Realms of Knowledge: Academic Departments in Secondary Schools* (London: Falmer Press, 1994).

3. Etienne Wenger, *Communities of Practice: Learning, Meaning, and Identity* (Cambridge, England: Cambridge University Press, 1998).

4. Lave, *Cognition in Practice*.

4/Taming the Monster: New Forms of Testing
That Encourage Learning

1. Alfie Kohn, *The Case Against Standardized Testing: Raising the Scores, Ruining the Schools* (Portsmouth, NH: Heinemann, 2000), 1.

2. Paul Black and Dylan Wiliam, *Inside the Black Box: Raising Standards through Classroom Assessment* (London: Department of Education and Professional Studies, King's College, 1998).

3. Paul Black, Chris Harrison, Clare Lee, Bethan Marshall, and Dylan Wiliam, *Assessment for Learning: Putting It into Practice* (Berkshire, England: Open University, 2003).

4. Phyllis Rosser, *The SAT Gender Gap: ETS Responds* (Washington, DC: Center for Women Policy Studies, 1992).

5. D. Pope, *Doing School: "Successful" Students' Experiences of the High School Curriculum* (Stanford, CA: Stanford University, 1999).

6. L. Bond, "My Child Doesn't Test Well," Carnegie Conversations (Stanford, CA: The Carnegie Foundation for the Advancement of Teaching, 2007).
7. Michael G. Gunzenhauser, "High-Stakes Testing and the Default Philosophy of Education," *Theory into Practice* 42, no. 1 (2003): 51–58.
8. Kohn, *The Case Against Standardized Testing*, 30.
9. Jo Boaler and Megan Staples, "Creating Mathematical Futures through an Equitable Teaching Approach: The Case of Railside School," *Teachers College Record* 110, no. 3 (2008): 608–45. Cited in the case of *Parents v. The Seattle Court District* (nos. 05-908 and 05-915). Available for download at www.youcubed.stanford.edu.
10. Jo Boaler, "When Learning No Longer Matters: Standardized Testing and the Creation of Inequality," *Phi Delta Kappan* 84, no. 7 (2003): 502–06. Available for download at www.youcubed.stanford.edu.
11. Barbara Y. White and John R. Frederiksen, "Inquiry, Modeling, and Metacognition: Making Science Accessible to All Students," *Cognition and Instruction* 16, no. 1 (1998): 3–118.
12. Black and Wiliam, *Inside the Black Box*.
13. Claude M. Steele, "A Threat in the Air: How Stereotypes Shape Intellectual Identity and Performance," *American Psychologist* 52, no. 6 (1997): 613–29.
14. Diane Reay and Dylan Wiliam, "'I'll Be a Nothing: Structure, Agency and the Construction of Identity Through Assessment," *British Educational Research Journal* 25, no. 3 (1999): 345–46.
15. Avraham N. Kluger and Angelo DeNisi, "The Effects of Feedback Interventions on Performance: A Historical Review, a Meta-Analysis, and a Preliminary Feedback Intervention Theory," *Psychological Bulletin* 119, no. 2 (1996): 254–84.
16. M. Deevers, *Linking Classroom Assessment Practices with Student Motivation in Mathematics*, paper presented at annual meeting of the American Educational Research Association held in San Francisco, 2006.
17. White and Frederiksen, "Inquiry, Modeling, and Metacognition."
18. Paul Black, Christine Harrison, Clare Lee, Bethan Marshall, and Dylan Wiliam, *Working Inside the Black Box: Assessment for Learning in the Classroom* (London: Department of Education and Professional Studies, King's College, 2004).
19. Maria Elawar and Lyn Corno, "A Factorial Experiment in Teachers' Written Feedback on Student Homework: Changing Teacher Behavior a Little Rather Than a Lot," *Journal of Educational Psychology* 77, no. 2 (1985): 162–73.
20. Ruth Butler, "Enhancing and Undermining Intrinsic Motivation: The Effects of Task-Involving and Ego-Involving Evaluation on Interest and Performance," *British Journal of Educational Psychology* 58, no. 1 (1998): 1–14.

21. Dylan Wiliam, "Keeping Learning on Track: Classroom Assessment and the Regulation of Learning," in Frank K. Lester Jr., ed., *Second Handbook of Mathematics Teaching and Learning* (Greenwich, CT: Information Age Publishing, 2007), 1085.

22. D. Royce Sadler, "Formative Assessment and the Design of Instructional Systems," *Instructional Science* 18 (1989): 119–44.

5/Stuck in the Slow Lane: How American Grouping Systems Perpetuate Low Achievement

1. James W. Stigler and James Hiebert, *The Teaching Gap: Best Ideas from the World's Teachers for Improving Education in the Classroom* (New York: Free Press, 1999).

2. Albert E. Beaton and Laura M. O'Dwyer, "Separating School, Classroom, and Student Variances and Their Relationship to Socioeconomic Status," in David F. Robitaille and Albert E. Beaton, eds., *Secondary Analysis of the TIMSS Data* (Dordrecht, Netherlands: Kluwer Academic Publishers, 2002).

3. Gerald Bracey, "Tracking, by Accident and by Design," *Phi Delta Kappan* 85, no. 4 (2003): 332–33.

4. Lisa Yiu, *Teaching Goals of Eighth Grade Mathematics Teachers: Case Study of Two Japanese Public Schools* (Stanford, CA: Stanford University, School of Education, 2001).

5. Carol Burris, Jay Heubert, and Henry Levin, "Accelerating Mathematics Achievement Using Heterogeneous Grouping," *American Educational Research Journal* 43, no. 1 (2006): 103–34.

6. Andrew C. Porter and Associates, *Reform of High School Mathematics and Science and Opportunity to Learn* (New Brunswick, NJ: Consortium for Policy Research in Education, 1994).

7. Robert Rosenthal and Lenore Jacobson, *Pygmalion in the Classroom: Teacher Expectation and Pupils' Intellectual Development* (New York: Holt, Rinehart and Winston, 1968).

8. Jo Boaler, Dylan Wiliam, and Margaret Brown, "Students' Experiences of Ability Grouping—Disaffection, Polarisation and the Construction of Failure," *British Educational Research Journal* 26, no. 5 (2000): 631–48.

9. PISA, Learning from Tomorrow's World.

10. Dylan Wiliam and Hannah Bartholomew, "It's Not Which School but Which Set You're In That Matters: The Influence of Ability-Grouping Practices on Student Progress in Mathematics," *British Educational Research Journal* 30, no. 2 (2004): 279–93.

11. Jeannie Oakes, *Keeping Track. How Schools Structure Inequality* (New Haven, CT: Yale University, 2005), 217.

12. Ibid., 218.

13. Steve Olson, *Count Down: The Race for Beautiful Solutions at the International Mathematical Olympiad* (New York: Houghton Mifflin, 2005), 48–49.
14. Jo Boaler, "The 'Psychological Prison' from Which They Never Escaped: The Role of Ability Grouping in Reproducing Social Class Inequalities," *Forum* 47, nos. 2, 3 (2006): 135–44.
15. Anabelle Dixon, Editorial, *Forum* 44, no. 1 (2002): 1.

6/Paying the Price for Sugar and Spice:
How Girls and Women Are Kept Out of Math and Science

1. Chi-squared = 16.96, n =163, 4 d.f., $p < 0.001$.
2. Anat Zohar and David Sela, "Her Physics, His Physics: Gender Issues in Israeli Advanced Placement Physics Classes," *International Journal of Science Education* 25, no. 2 (2003): 261.
3. Carol Gilligan, *In a Different Voice: Psychological Theory and Women's Development* (Cambridge, MA: Harvard University, 1982).
4. Mary F. Belenky, Blythe M. Clinchy, Nancy R. Goldberger, and Jill M. Tarule, *Women's Ways of Knowing: The Development of Self, Voice, and Mind* (New York: Basic Books, 1986), 214–29.
5. Janet S. Hyde, Elizabeth Fennema, and Susan Lamon, "Gender Differences in Mathematics Performance: A Meta-Analysis," *Psychological Bulletin* 107, no. 2, (1990): 139–55.
6. They recorded an effect size of only +0.15 standard deviations.
7. Rosser, *SAT Gender Gap*.
8. The examination was then called the O-level.
9. In 2003–2004, for example, 51 percent of the A, B, and C grades went to girls.
10. Joanne R. Becker, "Differential Treatment of Females and Males in Mathematics Class." *Journal for Research in Mathematics Education* 12, no. 1 (1981): 40–53.
11. Abbe H. Herzig, "Becoming Mathematicians: Women and Students of Color Choosing and Leaving Doctoral Mathematics," *Review of Educational Research* 74, no. 2 (2004): 171–214.
12. Abbe H. Herzig, "'Slaughtering This Beautiful Math': Graduate Women Choosing and Leaving Mathematics," *Gender and Education* 16, no. 3 (2004): 379–95.
13. Michele Cohen, "A Habit of Healthy Idleness: Boys' Underachievement in Historical Perspective," in Debbie Epstein, Jannette Elwood, Valerie Hey, and Janet Maw, eds., *Failing Boys?: Issues in Gender and Achievement* (Buckingham, England: Open University, 1999), 24.
14. John Bennett, cited in ibid., 25.

15. Pat Rogers and Gabriele Kaiser, eds., *Equity in Mathematics Education: Influences of Feminism and Culture* (London: Falmer, 1995).

7/Key Strategies and Ways of Working

1. Eddie Gray and David Tall, "Duality, Ambiguity, and Flexibility: A 'Proceptual' View of Simple Arithmetic," *Journal for Research in Mathematics Education* 25, no. 2 (1994): 116–40.
2. William P. Thurston, "Mathematical Education," *Notices of the American Mathematical Society* 37 (1990): 844–50.
3. T. A. Beck, "Are There Any Questions? One Teacher's View of Students and Their Questions in a Fourth-Grade Classroom," *Teaching and Teacher Education* 14, no. 8 (1998): 871–86.
4. Thomas L. Good, Ricky L. Slavings, Kathleen H. Harel, and Hugh Emerson, "Student Passivity: A Study of Question Asking in K–12 Classrooms," *Sociology of Education* 60 (1987), 181–99.
5. Boaler and Staples, "Creating Mathematical Futures."
6. James G. Greeno, "Number Sense as Situated Knowing in a Conceptual Domain," *Journal for Research in Mathematics Education* 22, no. 3 (1991): 170–218.
7. Jo Boaler and Cathy Humphreys, *Connecting Mathematical Ideas: Middle School Video Cases to Support Teaching and Learning* (Portsmouth, NH: Heinemann, 2005).
8. D. L. Ball, "With an Eye on the Mathematical Horizon: Dilemmas of Teaching Elementary Mathematics," *The Elementary School Journal* 93, no. 4 (1993): 373–97.
9. Magdalene Lampert, *Teaching Problems and the Problems of Teaching* (New Haven, CT: Yale University, 2001).
10. Cathy Humphreys and R. Parker, *Making Number Talks Matter. Developing Mathematical Practices and Deepening Understanding, Grades 4–10* (Portland, ME: Stenhouse, 2015).
11. Mathematics Education Collaborative, http://www.mec-math.org.
12. I am grateful to Emily Shahan for her in-depth analysis and descriptions of Jorge's experiences.
13. I am grateful to Tesha Sengupta-Irving for her in-depth analysis and descriptions of Alonzo's experiences.

8/Giving Children the Best Mathematical Start:
Activities and Advice for Parents
1. Nicholas Fiori, *In Search of Meaningful Mathematics: The Role of Aesthetic Choice* (doctoral dissertation, Stanford University, 2007).

2. M. Beth Casey, Ronald L. Nuttall, and Elizabeth Pezaris, "Mediators of Gender Differences in Mathematics College Entrance Test Scores: A Comparison of Spatial Skills with Internalized Beliefs and Anxieties," *Developmental Psychology* 33, no. 4 (1997): 669–80.

3. Eleanor Duckworth, *"The Having of Wonderful Ideas" and Other Essays on Teaching and Learning* (New York: Teachers College, 1996).

4. Allison M. Ryan and Helen Patrick, "The Classroom Social Environment and Changes in Adolescents' Motivation and Engagement During Middle School," *American Educational Research Journal* 38, no. 2 (2001): 437–60.

5. Jacquelynne S. Eccles, Allan Wigfield, Carol Midgley, David Reuman, Douglas Mac Iver, and Harriet Feldlaufer, "Negative Effects of Traditional Middle Schools on Students' Motivation," *The Elementary School Journal* 93, no. 5 (1993): 553–74.

6. Deborah Stipek and Kathy Seal, *Motivated Minds: Raising Children to Love Learning* (New York: Henry Holt, 2001).

7. M. Frank, "Problem Solving and Mathematical Beliefs," *Arithmetic Teacher* 35, no. 5 (1988): 32–34.

8. Joe Garofalo, "Beliefs and Their Influence on Mathematical Performances," *Mathematics Teacher* 82, no. 7 (1989): 502–5.

9. Flannery, *In Code.*

10. Ibid., 8.

11. Ibid.

12. Kenschaft, *Math Power,* 50.

13. Boaler and Humphreys, *Connecting Mathematical Ideas.*

14. Kenschaft, *Math Power,* 51.

15. Jacquelynne S. Eccles and Janis E. Jacobs, "Social Forces Shape Math Attitudes and Performance," *Signs: Journal of Women in Culture and Society* 11, no. 21 (1986): 367–80.

16. Sian L. Beilock, Lauren E. Holt, Catherine A. Kulp, and Thomas H. Carr, "More on the Fragility of Performance: Choking Under Pressure in Mathematical Problem Solving," *Journal of Experimental Psychology* 133, no. 4 (2004): 584–600.

17. Gray and Tall, "Duality, Ambiguity, and Flexibility."

9/Moving to a More Positive Future

1. See for example: Jeffrey E. Froyd, "Evidence for the Efficacy of Student-active Learning Pedagogies," April 22, 2007, http://trc.virginia.edu/wp-content/uploads/2013/07/Evidence-for-Efficacy-Froyd.pdf.

2. Carol S. Dweck, "The Perils and Promises of Praise," *Educational Leadership* 65, no. 2 (2007): 34–39.

3. Lisa S. Blackwell, Kali H. Trzesniewski, and Carol S. Dweck, "Implicit Theories of Intelligence Predict Achievement Across an Adolescent Transition: A Longitudinal Study and an Intervention," *Child Development* 78, no. 1 (2007): 246–63.

4. Eccles and Jacobs, " Social Forces."

5. Joseph J. Williams, Tania Lombrozo, and Bob Rehder, "Why Does Explaining Help Learning? Insight from an Explanation Impairment Effect," in *Proceedings of the 32nd Annual Conference of the Cognitive Science Society* (2010): 2906–11.

6. Moser et al., "Mind Your Errors."

7. http://online.stanford.edu/course/how-to-learn-math-for-students-s14.

8. Carol S. Dweck, "Is Math a Gift?: Beliefs That Put Females at Risk," in Stephen J. Ceci and Wendy M. Williams, eds., *Why Aren't More Women in Science?: Top Researchers Debate the Evidence* (Washington, DC: American Psychological Association, 2007).

9. Jo Boaler, "Changing the Conversation About Girls and STEM," paper presented to the Council on Women and Girls at the White House, April 28, 2014. Available for download at www.youcubed.stanford.edu.

10. Alina Tugend, "The Problem with Praise," *Worth,* http://www.worth.com/index.php/component/content/article/4-live/2908-the-problem-with-praise.

11. Boaler and Humphreys, *Connecting Mathematical Ideas.*

12. http://scpd.stanford.edu/instanford/how-to-learn-math.jsp.

13. Ibid.

14. Gray and Tall, "Duality, Ambiguity, and Flexibility."

15. Silva and White, *Pathways to Improvement.*

16. Boaler, "Fluency Without Fear."

17. Boaler, "Research Suggests."

18. http://online.stanford.edu/course/how-to-learn-math-for-students-s14.

19. Flannery, *In Code.*

Index